本书由中国科学院数学与系统科学研究院资助出版

数学 24/7

商业中的数学

〔美〕雷·西蒙斯 著

刘 刚 译

科学出版社

北 京

图字：01-2015-5622号

内 容 简 介

商业中的数学是“数学生活”系列之一，内容涉及如何找兼职赚钱、如何计算个人所得税等方面，同时介绍了在创业时计算成本和利润、贷款和利息、商品折扣和销售额以及发给雇员的工资等，让青少年在学校学到的数学知识应用到商业中的多个方面中，让青少年进一步了解数学在日常生活中是如何运用的。

本书适合作为中小学生的课外辅导书，也可作为中小学生的兴趣读物。

图书在版编目（CIP）数据

商业中的数学/（美）雷·西蒙斯（Rae Simons）著；刘刚译.—北京:科学出版社,2018.5
（数学生活）
书名原文：Business Math
ISBN 978-7-03-056745-1

Ⅰ.①商… Ⅱ.①雷… ②刘… Ⅲ.①数学-青少年读物 Ⅳ.①01-49

中国版本图书馆CIP数据核字（2018）第046666号

责任编辑:胡庆家 / 责任校对:邹慧卿
责任印制:肖　兴 / 封面设计:陈　敬

科学出版社 出版
北京东黄城根北街16号
邮政编码：100717
http://www.sciencep.com

北京汇瑞嘉合文化发展有限公司 印刷
科学出版社发行　各地新华书店经销
*
2018年5月第　一　版　　开本:889×1194　1/16
2018年5月第一次印刷　　印张:4　1/4
字数:70 000

定价：98.00元（含2册）
（如有印装质量问题，我社负责调换）

目　　录

Contents

引言

你会如何定义数学？它也许不是你想象的那样简单。我们都知道数学和数字有关。我们常常认为它是科学，尤其是自然科学、工程和医药学的一部分，甚至是基础部分。谈及数学，大多数人会想到方程和黑板、公式和课本。

但其实数学远不止这些。例如，在公元前5世纪，古希腊雕刻家波留克列特斯曾经用数学雕刻出了“完美”的人体像。又例如，还记得列昂纳多·达·芬奇吗？他曾使用有着赏心悦目的尺寸的几何矩形——他称之为“黄金矩形”，创作出了著名的画作——蒙娜丽莎。

数学和艺术？是的！数学对包括医药和美术在内的诸多学科都至关重要。计数、计算、测量、对图形和物理运动的研究，这些都被融入到音乐与游戏、科学与建筑之中。事实上，作为一种描述我们周围世界的方式，数学形成于日常生活的需要。数学给我们提供了一种去理解真实世界的方法——继而用切实可行的途径来控制世界。

例如，当两个人合作建造一样东西时，他们肯定需要一种语言来讨论将要使用的材料和要建造的对象。但如果他们建造的过程中没有用到一个标尺，也不用任何方式告诉对方尺寸，甚至他们不能互相交流，那他们建造出来的东西会是什么样的呢？

事实上，即便没有察觉到，但我们确实每天都在使用数学。当我们购物、运动、查看时间、外出旅行、出差办事，甚至烹饪时都用到了数学。无论有没有意识到，我们在数不清的日常活动中用着数学。数学几乎每时每刻都在发生。

很多人都觉得自己讨厌数学。在我们的想象中，数学就是枯燥乏味的老教授做着无穷无尽的计算。我们会认为数学和实际生活没有关系；离开了数学课堂，在真实世界里我们再不用考虑与数学有关的事情了。

然而事实却是数学使我们生活各方面变得更好。不懂得基本的数学应用的人会遇到很多问题。例如，美联储发现，那些破产的人的负债是他们所得收入的1.5倍左右——换句话说，假设他们年收入是24000美元，那么平均负债是36000美元。懂得基本的减法，会使他们提前意识到风险从而避免破产。

作为一个成年人，无论你的职业是什么，都会或多或少地依赖于你的数学计算能力。没有数学技巧，你就无法成为科学家、护士、工程师或者计算机专家，就无法得到商学院学位，就无法成为一名服务生、一位建造师或收银员。

体育运动也需要数学。从得分到战术，都需要你理解数学——所以无论你是想在电视上看一场足球比赛，还是想在赛场上成为一流的运动员，数学技巧都会给你带来更好的体验。

还有计算机的使用。从农庄到工厂、从餐馆到理发店，如今所有的商家都至少拥有一台电脑。千兆字节、数据、电子表格、程序设计，这些都要求你对数学有一定的理解能力。当然，电脑会提供很多自动运算的数学函数，但你还得知道如何使用这些函数，你得理解电脑运行结果的含义。

这类数学是一种技能，但我们总是在需要做快速计算时才会意识到自己需要这种技能。于是，有时我们会抓耳挠腮，不知道如何将学校里学的数学应用在实际生活中。这套丛书将助你一马当先，让你提前练习数学在各种生活情境里的运用。这套丛书将会带你入门——但如果想掌握更多，你必须专心上数学课，认真完成作业，除此之外再无捷径。

但是，付出的这些努力会在之后的生活里——几乎每时每刻（24/7）——让你受益匪浅！

1
工作：赚钱

阿里刚刚得到他的第一份工作，他想在暑期为上大学赚些学费。阿里向体育用品商店申请工作，随后被足球用品部门聘用。他热爱足球，迫不及待地想与客户谈论足球。

对于这项新工作，阿里有很多事情要做。他需要帮助顾客挑选商品，并推荐合适的商品，进货、上货，有时还帮着收银。

最令人激动的事情当然是阿里赚到的第一笔钱。他每小时赚9美元，整个暑假他都在努力工作，很快就赚了很多钱，那么他到底赚了多少呢？

阿里每周工作30小时。他知道自己每小时的工资是9美元，因此他可以很快算出每周可以赚多少钱：

$$9美元/小时 \times 30小时/周 = 270美元/周$$

那么整个夏天他赚了多少钱呢？为了找到答案，首先需要计算出暑假有多少周，一共有9周，其中1周要用于度假。因此

1. 270美元/周 × 8周 =

事实上，阿里通过临时照顾妹妹还可以额外赚到一些钱。妈妈为了说服他照看妹妹，愿意每小时支付10美元，这本来是要支付给他朋友的看护费。到目前为止，他最后一个月用来临时照看妹妹的时间为

第一周：2小时
第二周：4小时
第三周：0小时
第四周：2小时

阿里很想知道他平均每周照看妹妹需要多少时间，这样就可以推算出暑期剩下的时间可以赚多少钱。将各周小时数加起来，除以周数。

2. 阿里平均每周照看妹妹几小时？

现在阿里可以使用平均数来计算暑期照看妹妹赚多少钱。

3. 阿里照看妹妹一共赚了多少钱？记住，他的暑期有8周。

4. 阿里整个暑期一共赚了多少钱？

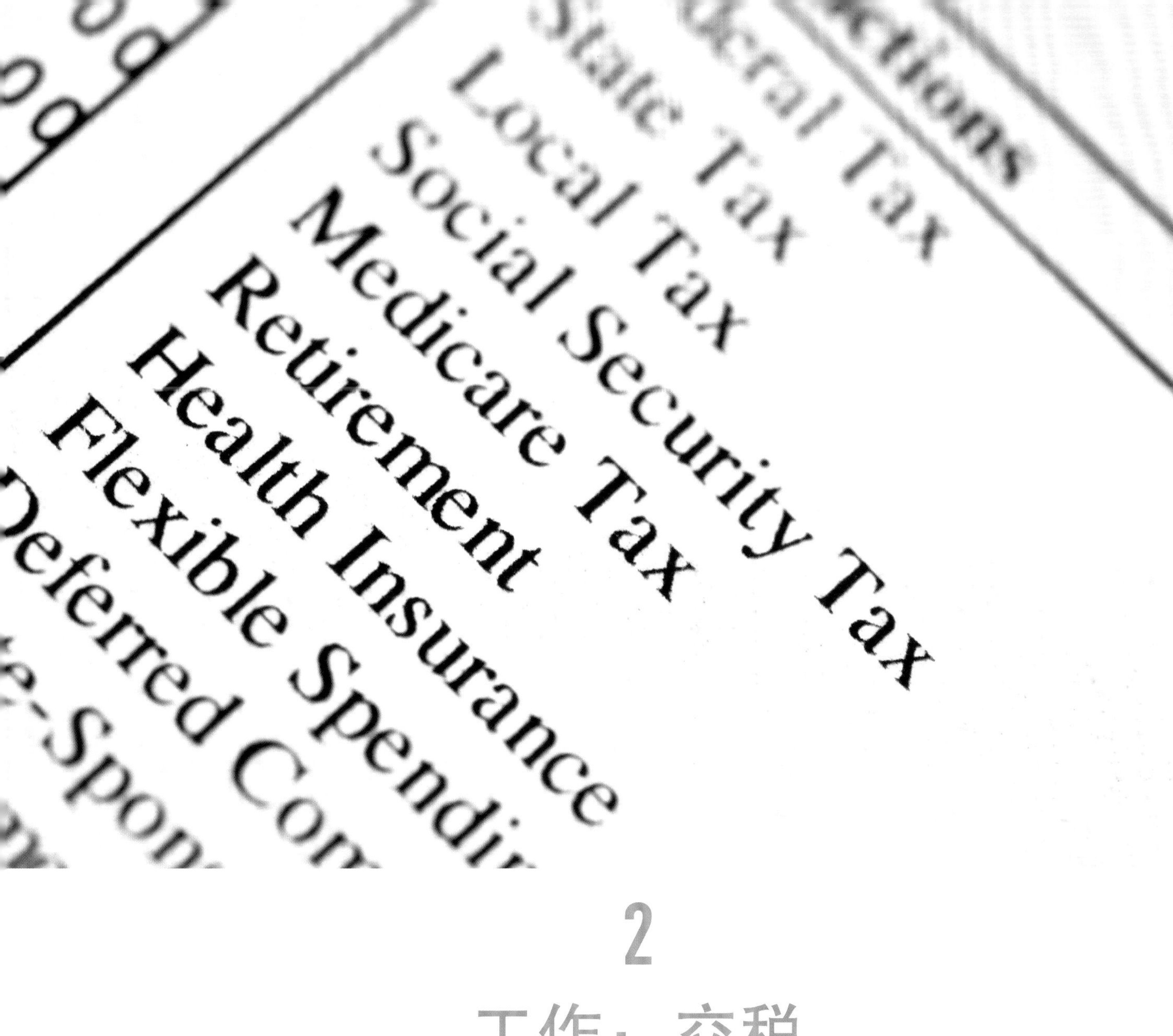

2
工作：交税

尽管阿里每周工作30小时，每小时赚9美元，但是他一周的收入却不是270美元，到手的只有206.14美元。这是怎么回事呢？

剩下的那些钱交税了。政府向工作中获得收入的人征税，这种税称为个人所得税。政府的税收用于警察、学校、道路、桥梁等的支出。缴纳个人所得税，实际上就是向每天都在使用的这些公共设施及服务付费。

阿里在工资条的下方看到纳税情况。他看到了四条纳税信息：联邦所得税、州所得税、社会保障税和医疗保险税。将来有一天，如果阿里想拥有自己的体育用品商店，他需要花费更多精力了解税费，因为商店老板需要支付物业(商场和土地)、营业收入和雇员方面的税费。但是目前，他只需要关注自己交的所得税。

阿里有以下需要支付的税种及税率：

联邦税：10%
州税：4%
社会保障税：6.2%
医疗保险税：1.45%

百分比是100份中的多少份，因此40%就是100份中的40份。问题在于阿里的工资超过100美元。他挣270美元，工资的40%当然超过40块。

你可以利用交叉相乘法来计算纳税额。比如联邦税：

$$\frac{10}{100}=\frac{X}{270}$$

$$100\times X=10\times 270$$
$$100\times X=2700$$
$$X=2700\div 100$$
$$X=27$$

你也可以通过向左移动两位小数点变百分比为十进制小数。现在，10%变为0.10。然后，乘以这个小数。对于联邦所得税，得到的答案和之前一样：

0.10 × 270美元=27美元

对于剩下的三种税，使用同样的方法来计算：

1. 阿里要支付多少州所得税？

2. 阿里要支付多少社会保障税？

3. 阿里要支付多少医疗保险税？

4. 阿里在商店每周赚206.14美元（税后），那么整个暑期他税后能赚多少钱？

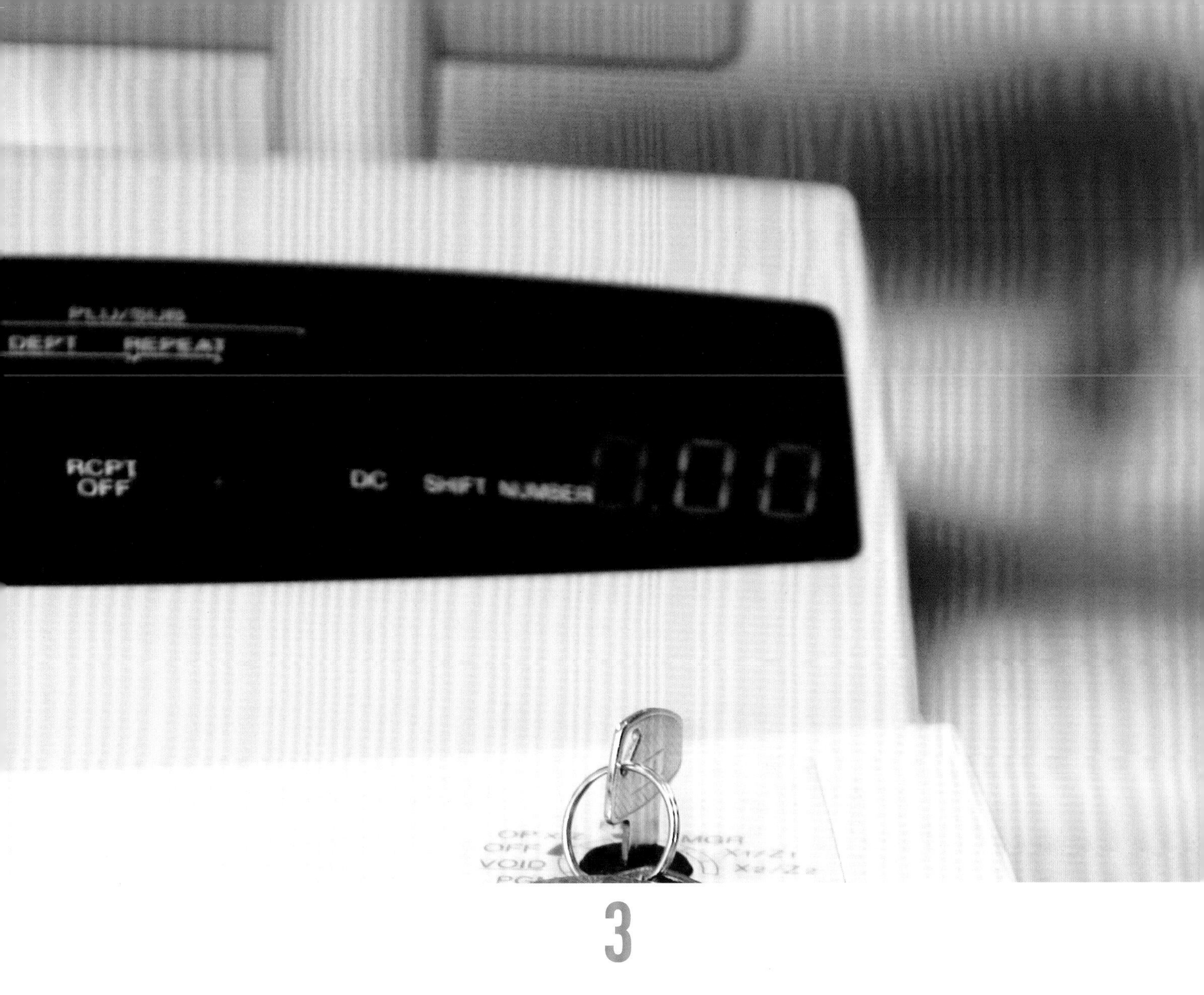

3

工作：找零

阿里的部分工作是收银。通常情况下这项工作不是很辛苦，因为阿里喜欢和人交流。他扫描商品上的条形码，收银机累加总金额。然后客户现金付款（或借记卡，或信用卡）。

他找零。有一天，收银机坏了，但阿里的老板却不打算关店，因为收银员仍然可以打开收银抽屉拿钱找零。

现在，阿里必须使用数学知识来计算顾客需要支付的钱，以及给顾客找零，速度还要足够快。

排在阿里收银台前的第一位顾客只买了一件东西：足球。一共花了14.99美元，顾客给了一张20美元的钞票。

1. 阿里应该给顾客找多少零钱？

下一个客户的货单要更复杂一些，阿里需要累加所有购买物品的钱款：

运动鞋：51.50美元
网球拍：129.00美元
网　球：2.59美元

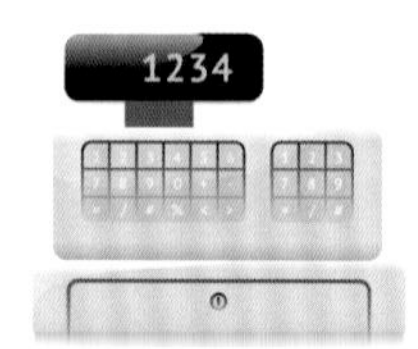

2. 顾客需要付多少钱？

这个顾客没有借记卡或信用卡，准备用现金支付。她递给阿里一张50美元的钞票，六张20美元的钞票，三张1美元的钞票和25美分。阿里从来没有见过这么多现金！

3. 顾客一共给了阿里多少钱？她给的够吗？

阿里算了一下，然后告诉她钱不够。顾客又给了阿里一张20美元的钞票，然后要求找零。阿里心算了一下。

4. 阿里应该给顾客找零多少钱？

4
工作：支票账户

阿里在这个暑期攒了这么多钱，他想要开一个支票账户。他已经有了一个用来存大学学费的储蓄账户。

支票账户用来保管所有你将要花的钱。把钱存入支票账户要比放在卧室或者衣服口袋里安全得多。阿里发现虽然拿着现金会很开心，但容易丢失。

阿里去银行开通了一个支票账户。银行职员告诉他开支票账户的条件，保证账户最低余额不少于25美元，否则将收管理费。他的借记卡也要求每月至少使用两次以保持账户的活跃状态。否则，银行可能会冻结账户。他认为可以做到这些，所以他申请支票账户，并存入25美元。现在他要做的就是关注他的钱，并且不能花太多！下一页将介绍阿里如何管理他的支票账户。

阿里放在家里的现金总共有48.83美元，全部存入支票账户。算上以前存入的25美元，现在他有73.83美元。

接下来的工作日，阿里收到了第一张工资支票（206.14美元）。他去银行存了一半到支票账户，另一半存入储蓄账户攒学费。

1. 现在他的支票账户有多少钱？

阿里真的很高兴有了更多的钱可以支配。现在他可以买一个期待很久的新的MP3播放器，需要花140美元。

2. 他有足够的钱吗？如果够的话，还剩多少？

3. 剩下的钱超过25美元吗？这是支票账户免收服务费的最低限额。

阿里与朋友外出，准备买冰激凌，他不清楚支票账户里剩多少钱，但他认为够给自己买冰激凌了。他的朋友路易莎希望他帮忙买一个圣代冰激凌，回头还他钱。阿里不想拒绝她，于是一共支付了12.5美元用于购买冰激凌和甜点。

4. 他的支票账户里还有能满足最低限额的钱吗？如果不能，缺多少？

5

工作：对账

阿里正在用功学习理财！由于他花了很多钱导致支票账户余额不足，支付了30美元的管理费，他决定将来花钱时更谨慎一些。

开户后，阿里有了一张借记卡和一本支票簿。因为不会用，到目前为止他还没有使用支票簿。碰巧埃利亚斯叔叔来了，看到桌子上阿里的支票簿，于是教他如何开支票和对账。

埃利亚斯叔叔告诉阿里可以存钱查看并核对支票账户余额，在网上能看到每一笔支出和收入。任何时候，他都应该关注支票簿上的收支情况。这样，如果银行或商店出错，阿里也可以获知，并确保没有多付钱。下面你可以看看埃利亚斯叔叔怎么教阿里的。

首先，埃利亚斯叔叔告诉阿里整理所有商场收据和银行存根。下面是他收集的情况：

10/27—电子游戏机，35.99美元
10/29—工资，206.14美元
11/3—电影票，11美元
11/7—看护费，30美元

现在把阿里收集的所有单据填到下面的图表内。购买在支出列，存钱在收入列。

日期	描述	支出/美元	收入/美元	余额/美元
10/27	电子游戏机	35.99	—	74.37
10/29	工资	—	206.14	280.51

1. 阿里最后还剩多少钱？

2. 如果阿里开一张300美元的支票，他支票账户的钱够吗？如果够的话，他还剩下多少钱？如果不够，他还缺多少钱？

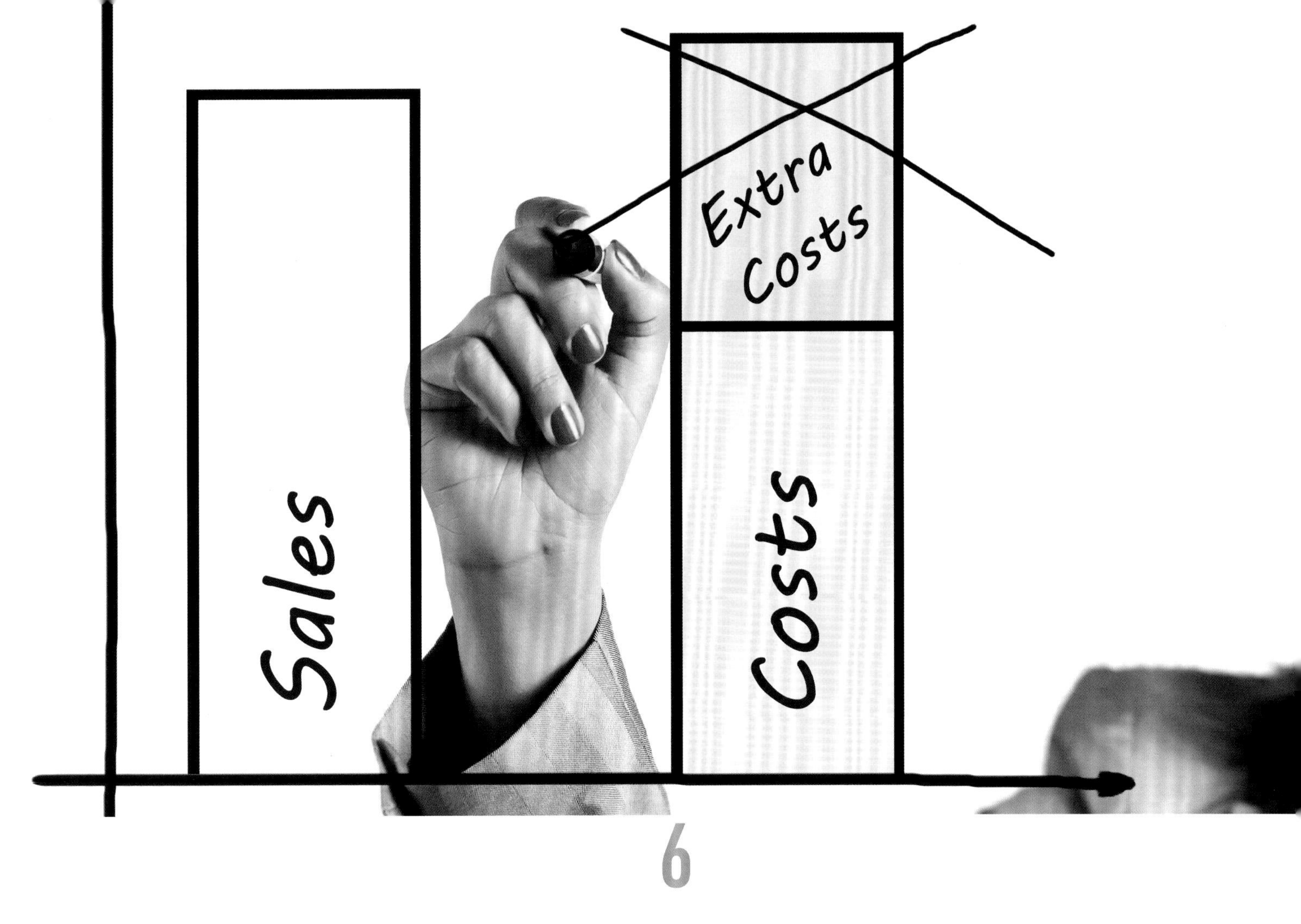

6

创业：利润和成本

阿里的姐姐伊娃想尝试创业。她一直想拥有自己的事业，如今终于开始了她的梦想。因为她喜欢栽培植物，所以伊娃从园艺方面进行创业。

伊娃做了很多计划。她计划从大家院子里的景观或走廊里的盆景生意开始。她将提供植物、园艺工具和其他用品，也将教人们如何照料他们的花园。

然而，她不确定应该向客户收取多少费用。她不希望收费太高，以至于顾客无法承担费用而不光顾她的生意。但她又想要赚钱，即除去成本后能有所获利。她怎么去算呢？下面你将看到她如何确定收费标准。

首先，伊娃列出了工作时需要花的钱。她为两种类型的景观提供不同的价格：一种是花园景观价格，一种是对于院子空间不够或者压根没有院子的顾客提供走廊景观价格。

花园景观：
植物：150美元
堆肥：40美元
化肥：6美元

走廊景观：
植物：115美元
盆：90美元
泥土：40美元
堆肥：10美元
化肥：6美元

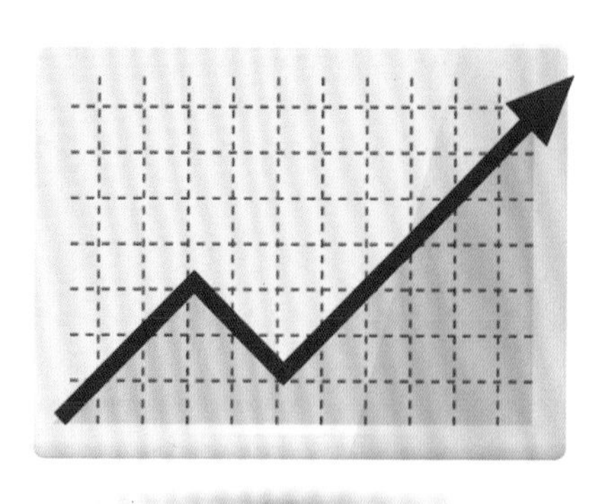

1. 每种景观的总成本分别是多少？

如果她仅仅只收取建造景观的成本，那样是赚不到钱的！她还需要计算时间成本。开创事业和建造景观所付出的心血，这些都是有价值的。她认为工作一小时价值20美元，因为她认为付出的劳动和努力值这么多。

2. 如果她构建一个花园景观需平均工作8小时，走廊景观需工作3个小时，每小时20美元，那么她一共能赚多少钱？

现在你可以将伊娃的时间成本和建造景观的成本加在一起，计算出向客户收取的费用。此外，伊娃还想多收点费用，她认为客户会愿意多付一些。因此她对每个景观多收30美元。那么对一个花园景观，伊娃应该收取顾客多少钱？

196美元＋160美元＋30美元＝386美元

3. 走廊景观应该收费多少？

现在你可以算出伊娃的利润。你知道每个花园材料需要多少钱，以及她的收费是多少。那么花园景观的利润就是收费减去成本：

386美元－196美元＝190美元

4. 伊娃建造走廊景观可以获得多少利润？

5. 哪种类型景观她应该做多一些？为什么？

7
创业：办公面积

为开展她的花园生意，伊娃知道她要租一间小的办公室。这样她才能会见客户、储存文件，以及在办公室的电脑上工作。如果生意足够好，她还得雇人在办公室工作。

她逛了一些办公室，她清楚只能负担一个小的。考虑到工作主要在户外，一个小的办公室也够用了。在房租方面，一个月最多花500美元。那么她能找到这样的办公室吗？下面我们将帮她决定。

伊娃看的第一个办公室的租金是每年30美元/平方英尺，这个房间长30英尺，宽15英尺。那么她能租得起吗？

这里是面积的计算公式：

面积(平方英尺)＝长×宽

面积＝30×15

1. 办公室的面积＝________平方英尺

将办公室面积乘以租金，除以12个月，这样得到每月需要多少租金。

2. 这在伊娃的预算内吗？

伊娃看的第二个办公室是每年18美元/平方英尺，面积是330平方英尺。

3. 她能负担得起这个吗？为什么？

伊娃的办公室至少要能容纳两张书桌、三个文件柜、一张办公桌和几张椅子。桌子大小是3.5英尺×1.5英尺，文件柜为1英尺×1英尺，桌子和椅子占用5英尺×4英尺的空间。另外她还需要活动空间，至少100平方英尺。那么这个办公室有足够空间吗？

桌子面积＝2×(3.5×1.5)

4. 桌子的面积＝

5. 文件柜的面积＝

6. 桌子和椅子的面积＝

7. 现在所有面积相加。别忘了还要考虑活动空间。

8. 办公室里有足够的空间吗？如果够，它还剩多少额外空间？如果不够，它还需要多少空间？

8

创业：贷款

伊娃没有足够的钱来开始自己的事业。比如，她必须买手推车、水管和其他景观工具。她还必须购买植物和支付前几个月的办公室房租。她认为生意一旦开始，就可以赚到很多钱。但是，首先她得开始，而且生意需要用钱来赚钱。

幸运地是，伊娃可以从银行获得贷款，贷款是借来的钱。一旦景观生意开始赚钱，伊娃就能够向银行还贷。

伊娃去银行咨询贷款的事情。她向一个名叫米格尔的银行雇员咨询她可以贷多少钱。米格尔给她算了一下。下面是米格尔帮她算的结果。

伊娃首先得算出需要多少钱，这样就可以算出她应该贷多少。

这里列出了伊娃开始生意所需要的各种费用：

第一个月办公室租金：495美元
手推车：45.50美元
软管：19.99美元
干草叉：32.99美元
最初几个花园的材料：600美元
办公用品：35美元
2个桌子：80美元/个
2个椅子：48.75美元/个
3个文件柜：17.50美元/个
办公桌和椅子：160.00美元
打印机：200.00美元
广告：50美元
货车油钱：60美元
营业执照：150美元

1. 伊娃一共需要多少钱？

然而，伊娃并不需要借那么多钱。为了创业，她存了一些钱，虽然她花了一些，但还是有结余可以用于创业起初的开支。她还剩600美元。

伊娃创业需要的资金减去她存的钱就是需要贷的钱。

2. 伊娃要贷多少钱？

9
创业：利息

伊娃在银行的时候，米格尔告诉过她关于贷款利息的事情。利息是为获得贷款而产生的额外费用。你不需要马上支付利息——在开始还款时利息加到贷款本金上了。

高利率意味着伊娃将需要为她的贷款支付更多利息，低利率意味着少一些。

米格尔说，银行贷款有复利。伊娃将贷款1560美元。米格尔告诉她银行利率是7.5%，而且她需要在两年内还清贷款。她的第一笔利息是基于最初1560美元的贷款。第二年，她将支付更多的利息，包括贷款本金的利息，加上她欠的第一年利息的利息。银行把这笔第一年的利息作为第二笔小额贷款，所以也要收取利息。这就是“复利”的意思（注：两年内付清，如果在一年内付清就没有复利，否则将产生复利）。

你可以看到利息增加的速度很快！下面我们将计算伊娃的贷款需要支付多少利息。

这里的数学公式用来计算利息后会有多少钱：

$$总额=P(1+r/n)^{nt}$$

公式看起来很复杂，但如果知道了这些字母的意思，公式也不是很难。

P=你贷了多少钱，称为本金
r=利率，银行会给出具体数字；r经常是一个小数
n=利息计算的次数，可能是一月一次或一年四次，或一年一次
t=年数

nt是指数，表示一个数乘以本身的次数。2^3是$2\times2\times2$，等于8。计算器可以做指数运算。

伊娃的信息是：

$P=1560$美元
$r=0.075(=7.5\%)$
$n=12$，因为每月计息一次
$t=2$年

那么计算伊娃在两年含利息一共欠多少钱？

1. 总金额$=1560\times(1+0.075/12)^{12\times2}$

   总金额=

2. 总额中多少是利息？

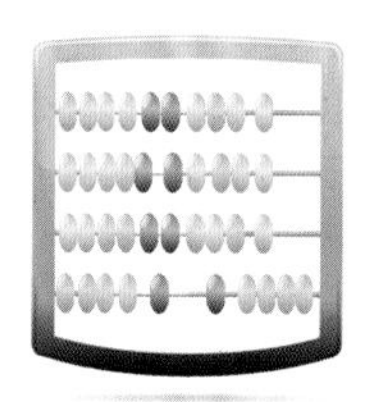

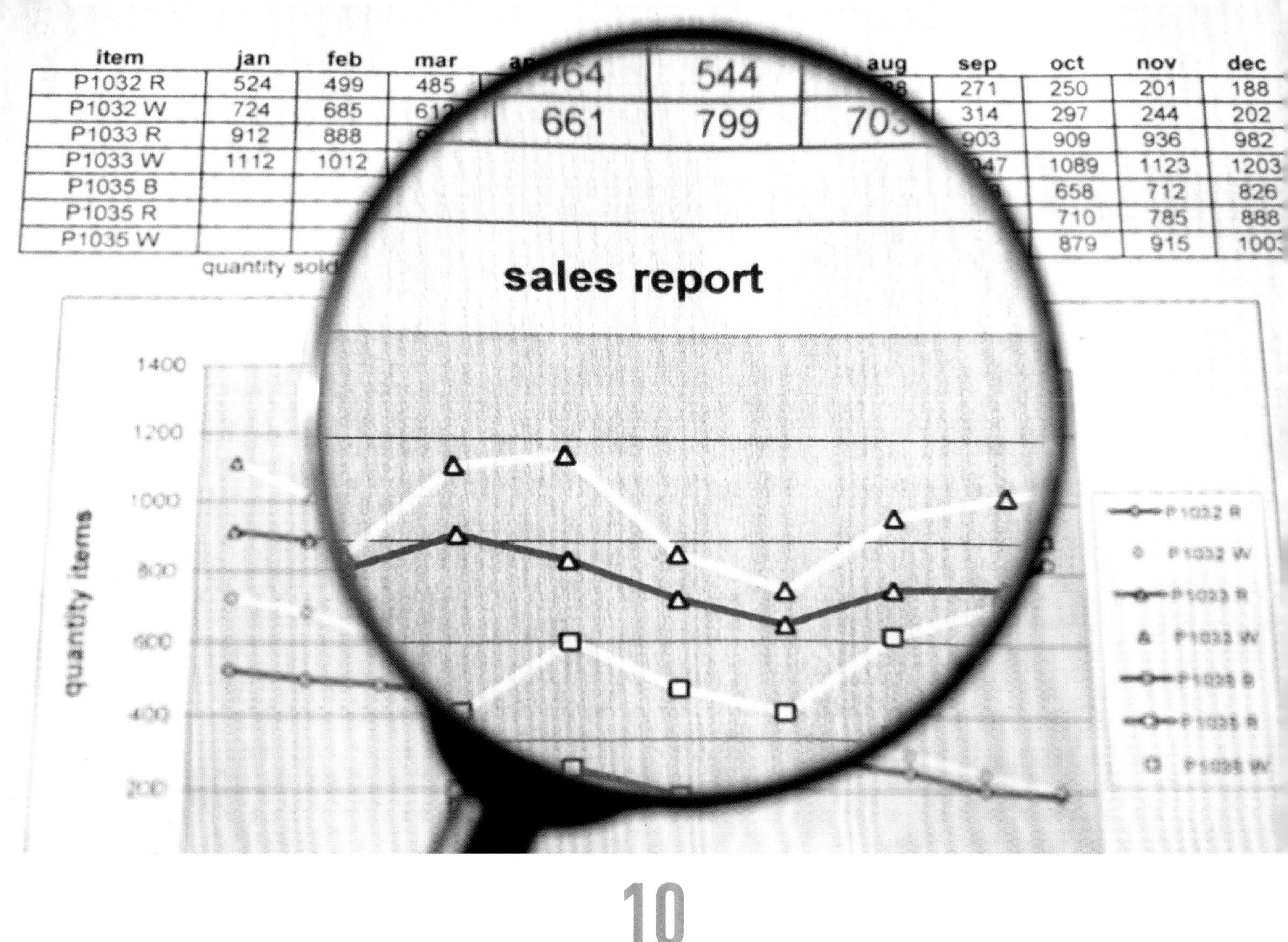

10

创业：销售额

现在伊娃的生意已经开始。她开始进行广告宣传，装饰办公室，并为第一个客户的花园种了植物。伊娃想要认真地记录她的生意，这样就可以准确地知道生意运转的状态。

伊娃特别关注的是销售额，也就是她向客户收取的费用总和。

顾客越多，她的生意越多，赚的钱也就越多。她第一个月的生意有些少，但在第二个月就有很多顾客了。

如果经营得不错，足以还贷，那么伊娃就能计算出她的利润。看看她的记录，计算销售额。

第一个月，伊娃有3个客户，做了2单花园景观和1单走廊景观生意。

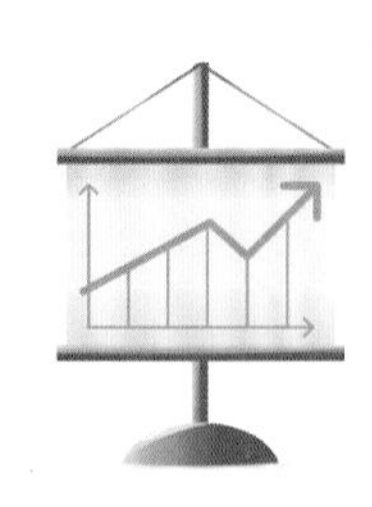

1. 她第一个月挣了多少钱？看看前面每个景观的成本是多少。

起初，伊娃是基于景观建造的成本来计算向客户收取的费用。然而，她忘了包含每月还贷的支出。为了在两年内还清贷款，她每月需还贷75美元。第一个月后，还必须支付房租和汽油钱。

她计算每个景观的建设成本，再加上第一个月的还贷支出。如果总额低于本月的收入，那么差额就是利润。

2. 伊娃有利润吗？如果有，是多少？如果没有，她还需要多赚多少钱？

第二个月，伊娃花了更多时间来推销。她最终做了5单花园景观和4单走廊景观生意。但这个月，她还必须支付办公室租金和60美元的油钱，再加上还贷和景观材料的支出。

3. 相较上月，伊娃的利润变多了还是变少了？

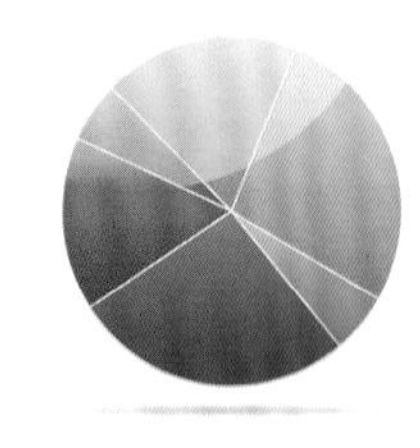

11
创业：订货

每次获得订单，伊娃都必须订购建造新花园的材料。她和客户一起商量他们想要什么和花园要建设成什么样。有些客户想要在院子里面搭建花园，也有客户想要在门外建灌木丛，还有客户想在走廊上种几盆蔬菜。

伊娃需要找出最好的供应商。她喜欢从这家供应商买花，从那家供应商买蔬菜幼苗，从这家供应商买花盆，从那家供应商买泥土。有时她也记不清这些东西是从哪个供应商买的，但是她知道保持联系很重要。

为确保到货时正确付款，伊娃每次都要检查订单。她已经发现过几次错误了，所以她知道值得花时间去检查。

通常伊娃填的订购单是这样的：
日期：5/16
订单号：12345678

货物描述	数量	单价	总计
南瓜苗	4	3.99美元/株	15.96美元
番茄苗	5	6.99美元/株	34.95美元
茄子苗	3	5.50美元/株	16.50美元
豌豆苗	10	2.59美元/株	25.90美元
萝卜种子	1	3美元/包	3美元
莴苣种子	1	3.50美元/包	3.50美元

总计：99.81美元
税(8%)：7.98美元
运送费：6.54美元
实际应付货款：114.33美元

她实际上获得了：

4株南瓜苗
4株番茄苗
0个茄子苗
10株豌豆苗
1包萝卜种子
0包莴苣种子

供应商将伊娃的订单搞得很乱！

1. 伊娃的订单与货车实际送来的货物之间的区别是什么？

2. 伊娃的订单和她实际收到的货物之间的价格差异是多少？

12

创业：折扣

伊娃正和她的弟弟阿里讨论她的生意。她已经做得很好了，但还想吸引更多的顾客。她已经做了很多推销工作，但是不确定还能做什么来说服更多的人雇她建造景观。

阿里有个很好的主意——他建议伊娃给顾客打折。这样，那些认为价格过高的顾客将觉得可以负担得起一个景观。

伊娃同意并马上付诸行动。下面我们将计算她应该打折多少。

伊娃将为建造景观提供一定比例的折扣。首先她考虑提供25%的折扣券。

对于计划订购花园景观的顾客，25%的折扣是多少钱呢？

你可以用几个不同的方式计算这25%。首先，可以把百分比化为分数：25%和1/4是一样的。因此，一个花园景观需花费386美元：

$$386美元 \div 4 = 96.50美元$$

下一步，从花园景观总的开支中减去打折优惠的费用：

$$386美元 - 96.50美元 = 289.50美元$$

你也可以将百分比转换成小数。这意味着25%就是0.25。然后原价乘以0.25，从原始成本减去这笔优惠的费用，会得到相同的答案：

$$0.25 \times 386美元 = 96.50美元$$
$$386美元 - 96.50美元 = 289.50美元$$

现在以走廊景观为例进行计算。

1. 如果客户得到25%的折扣，那么他们还要付多少钱？

伊娃只是想在盈利的前提下提供25%的折扣，这样她仍然有利润，并且签订的新客户还会向朋友们推荐她。

2. 如果提供25%的折扣，伊娃的花园景观还能盈利吗？如果是的话，折后利润是多少，假设不需要考虑每月还贷。翻回前面的几节找到你需要的数据。

3. 她的走廊景观赚钱吗？如果是的话，赚多少？

4. 根据你的答案，你觉得伊娃应该在花园景观还是在走廊景观上打25%的折扣？

13

创业：佣金

伊娃认为她弟弟阿里很有商业头脑。生意正有起色，顾客也比较多，因此她需要一点帮助。她问阿里是否愿意为她打工。

她解释说，报酬不像雇员那么固定。事实上，阿里需要努力向客户推销。每做成一单，他将赚到一些钱。这些钱是对他做成生意的奖励。工作越努力，他就挣的越多。

阿里仔细考虑了一下，决定要挣更多的钱。他返校后在体育用品店工作的时间减少了。给姐姐打工，工作时间比较灵活，还可以攒些学费和得到一些零用钱。

对于获得订单的员工，雇主一般支付固定的报酬，或者按销售额的比例支付，如5%或15%。下面展示伊娃是如何计算阿里的佣金的。

阿里每次向顾客推销一个花园景观，伊娃支付阿里一定数量的报酬，那么每个花园应该付他多少报酬？

她决定按销售额的一定比例支付阿里报酬。她只卖两种产品，所以计算支付多少不难计算。

首先，她试着按1%计算佣金。

花园景观销售额的1%是多少？用小数表达。

1. 0.01 × 386美元=

2. 你认为阿里是否会花费时间销售景观来赚取佣金？如果需要花费两小时做成一单，他是否会选择其他方式去赚更多的钱？

伊娃看到1%的佣金太少。现在考虑为花园景观支付10%的佣金：

3. 0.1 × 386美元 =

4. 走廊景观销售额的10%作为佣金是多少呢？

5. 对于打折25%后的花园景观，阿里可以赚多少钱？

10%的佣金看起来不错，阿里决定接受这份工作。

Now
HIRING

14

创业：员工

几周后，阿里为景观生意带来了7个新客户。但伊娃几乎没有时间来处理新业务，她的工作跟不上了。这时她有了一个新主意——她应该雇弟弟兼职。他能做销售，同样也可以帮着建景观。

一方面，雇佣一个员工需要花掉公司很多钱。另一方面，如果做不完工作，她的生意就不能发展。阿里已经从计件工作中获得近300美元的报酬，所以伊娃知道她至少需要支付几个小时的兼职费用。她计算了一下，看看一个星期可以雇佣弟弟多少个小时。

伊娃知道要提供至少与每小时在体育用品商店的收入相当的工资。否则，弟弟没理由离开那份工作。所以，她至少得给阿里9美元/小时。

1. 如果每周雇佣阿里10小时，每小时11美元，那么伊娃每周需要支付多少钱？

2. 与每周在体育用品店工作10小时相比，那么阿里为伊娃工作每周可以多赚多少钱？

现在伊娃决定是否按照每周10小时、每小时11美元雇佣她的弟弟。

阿里每月工作4周，所以她每月将付给他440美元。伊娃希望在员工的帮助下可以建造更多的景观。她估计公司至少每月可以多建6个景观。

计算她的利润区间，也就是最低利润到最高利润的范围。首先，计算多建造6个走廊景观的利润，这是最低的利润，是利润区间的下界。下面我们将具体计算。

3. 6 × 90美元 =

现在计算6个花园景观的利润，这是最高的利润，是区间的上界。

4. 区间的上界是多少？

5. 通过雇佣阿里，她可以获得利润的范围是多少？

6. 你认为值得去雇阿里吗？为什么？

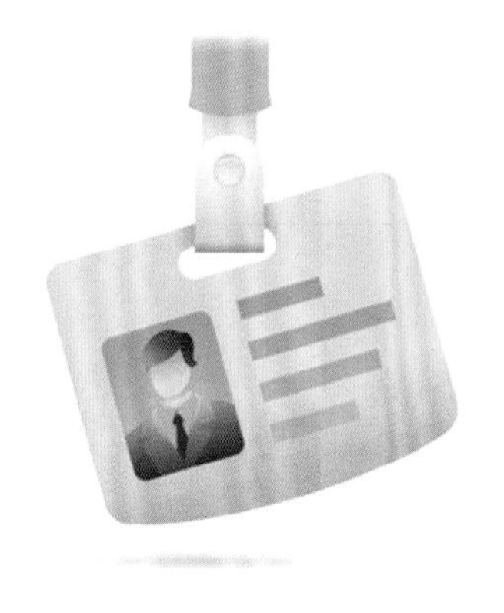

15
小 结

伊娃最终雇佣了阿里，而且她的生意确实更好了。伊娃的业务很成功！她从第一天创业开始经营景观生意就已经取得很大的成功。阿里作为雇员一路走来也取得了很大进步。

看看你能否记得阿里和伊娃从工作和创业中所用到的知识。

1. 如果阿里加薪到13美元/小时，且每周工作10个小时，阿里一周可以赚到多少钱？

2. 与之前的税率相同，阿里每周实际工资进账多少钱？

3. 阿里把每周薪水的35%存入支票账户，如果一分不花，四周后支票账户有多少钱？

4. 如果将奶奶给的生日礼物——一张50美元支票存进支票账户，那么阿里是否有足够的钱买一个200美元的电视。

5. 从供应商处购买的花盆突然涨价了，现在是5美元1个，一般她需要买5个。那么如果景观不涨价，她的利润将减少多少？

6. 伊娃的花园景观通常是8英尺×6英尺，那么花园的面积是多少平方英尺？

7. 如果其他因素不变，伊娃要在一年内付清贷款，那么她每月要还贷多少？

8. 伊娃提供了一个新的花园景观销售方案：每个花园景观的折扣是15%。那么一个花园景观是多少钱？

参考答案

1.

1. 2160美元
2. 2小时
3. 10美元/小时 × 2小时/周 × 8周 = 160美元
4. 2160美元 + 160美元 = 2320美元

2.

1. 10.80美元
2. 16.74美元
3. 3.92美元
4. (206.14美元/周 × 8周) + 160美元 = 1809.12美元

3.

1. 20美元 - 14.99美元 = 5.01美元
2. 51.50美元 + 129.00美元 +2.59美元 = 183.09美元
3. 不够，她给阿里173.25美元
4. 193.25美元 - 183.09美元 = 10.16美元

4.

1. 73.83美元 +(206.14美元/2) = 176.90美元
2. 是的，她将剩下36.9美元
3. 是
4. 不是，她有24.4美元，比需要的缺少0.6美元

5.

1. 299.51美元
2. 不够。他还需要25.49美元，其中包括支票账户中最少需保留的余额

日期	描述	取款/美元	存款/美元	收支平衡/美元
10/27	电子游戏机	35.99	—	74.37
10/29	工资	—	206.14	280.51
11/3	电影票	11	—	269.51
11/7	看护费	—	30	299.51

1. 花园景观: 196美元；走廊景观: 261美元
2. 花园景观: 20美元 × 8 = 160美元；走廊景观：20美元 × 3 = 60美元
3. 261美元 + 60美元 + 30美元 = 351美元
4. 351美元 - 261美元 = 90美元
5. 花园景观，因为利润更高

7.

1. 450
2. 不，办公室租金为1125美元/月，超出500美元/月的预算
3. 能，此时租金为495美元/月
4. 10.5平方英尺
5. 3平方英尺
6. 20平方英尺
7. 10.5 + 3 + 20 + 100 = 133.5平方英尺
8. 够，还剩196.5平方英尺

8.

1. 2158.48美元
2. 2158.48美元 - 600美元 = 1558.48美元

9.

1. 1811.62美元
2. 1811.62美元-1560美元=251.62美元

10.

1. (2×386美元)+351美元=1123美元
2. 有，她每月的利润是395美元
3. (5×386美元)+(4×351美元)=3334美元
   3334美元-(75美元+(5×196美元)+(4×261美元)+495美元+60美元)=680美元
   她本月的利润比上月多了285美元

11.

1. 共缺少1株番茄苗，3株茄子苗和1包莴苣种子
2. 6.99美元+16.50美元+3.50美元=26.99美元

12.

1. 263.25美元
2. 盈利，她仍然有93.50美元的利润
3. 盈利，不过只有2.25美元
4. 她应该关注在花园景观上打折，因为折后利润更多

13.

1. 3.86美元
2. 不会，运动品店或照看孩子收入更高
3. 38.60美元
4. 0.10×351美元=35.10美元
5. 0.10×289.50美元=28.95美元

14.

1. 11美元 × 10 = 110美元
2. 每周多赚20美元
3. 540美元
4. 1176美元
5. 540美元 - 440美元 = 100美元(下界), 1176美元 - 440美元 = 736美元(上界)
6. 值得, 因为无论如何生意将盈利

15.

1. 130美元
2. 13美元 + 5.20美元 + 8.06美元 + 1.89美元 = 28.15美元,
   130美元 - 28.15美元 = 101.85美元
3. 4 × (0.35 × 101.85美元) = 142.59美元
4. 不够, 他没有足够的钱去买电视
5. 5美元 × 5 = 25美元
6. 6英尺 × 8英尺 = 48平方英尺
7. 总数计算公式 = 1560美元 $(1 + 0.075/12)^{12 \times 1}$
   总数 = 1681.11美元
   1681.11/12月 = 140.09美元/月
8. 0.15 × 386美元 = 57.90美元, 386美元 - 57.90美元 = 328.10美元

INTRODUCTION

How would you define math? It's not as easy as you might think. We know math has to do with numbers. We often think of it as a part, if not the basis, for the sciences, especially natural science, engineering, and medicine. When we think of math, most of us imagine equations and blackboards, formulas and textbooks.

But math is actually far bigger than that. Think about examples like Polykleitos, the fifth-century Greek sculptor, who used math to sculpt the "perfect" male nude. Or remember Leonardo da Vinci? He used geometry—what he called "golden rectangles," rectangles whose dimensions were visually pleasing—to create his famous *Mona Lisa*.

Math and art? Yes, exactly! Mathematics is essential to disciplines as diverse as medicine and the fine arts. Counting, calculation, measurement, and the study of shapes and the motions of physical objects: all these are woven into music and games, science and architecture. In fact, math developed out of everyday necessity, as a way to talk about the world around us. Math gives us a way to perceive the real world—and then allows us to manipulate the world in practical ways.

For example, as soon as two people come together to build something, they need a language to talk about the materials they'll be working with and the object that they would like to build. Imagine trying to build something—anything—without a ruler, without any way of telling someone else a measurement, or even without being able to communicate what the thing will look like when it's done!

The truth is: We use math every day, even when we don't realize that we are. We use it when we go shopping, when we play sports, when we look at the clock, when we travel, when we run a business, and even when we cook. Whether we realize it or not, we use it in countless other ordinary activities as well. Math is pretty much a 24/7 activity!

And yet lots of us think we hate math. We imagine math as the practice of dusty, old college professors writing out calculations endlessly. We have this idea in our heads that math has nothing to do with real life, and we tell ourselves that it's something we don't need to worry about outside of math class, out there in the real world.

But here's the reality: Math helps us do better in many areas of life. Adults who don't understand basic math applications run into lots of problems. The Federal Reserve, for example, found that people who went bankrupt had an average of one and a half times more debt than their income—in other words, if they were making $24,000 per year, they had an average debt of $36,000. There's a basic subtraction problem there that should have told them they were in trouble long before they had to file for bankruptcy!

As an adult, your career—whatever it is—will depend in part on your ability to calculate mathematically. Without math skills, you won't be able to become a scientist or a nurse, an engineer or a computer specialist. You won't be able to get a business degree—or work as a waitress, a construction worker, or at a checkout counter.

Every kind of sport requires math too. From scoring to strategy, you need to understand math—so whether you want to watch a football game on television or become a first-class athlete yourself, math skills will improve your experience.

And then there's the world of computers. All businesses today—from farmers to factories, from restaurants to hair salons—have at least one computer. Gigabytes, data, spreadsheets, and programming all require math comprehension. Sure, there are a lot of automated math functions you can use on your computer, but you need to be able to understand how to use them, and you need to be able to understand the results.

This kind of math is a skill we realize we need only when we are in a situation where we are required to do a quick calculation. Then we sometimes end up scratching our heads, not quite sure how to apply the math we learned in school to the real-life scenario. The books in this series will give you practice applying math to real-life situations, so that you can be ahead of the game. They'll get you started—but to learn more, you'll have to pay attention in math class and do your homework. There's no way around that.

But for the rest of your life—pretty much 24/7—you'll be glad you did!

1 WORKING: EARNING MONEY

Ari just got his first job. He's out of school for the summer, and he wants to earn money for college. Ari applied for a job at a sporting goods store, and he got hired to work in the soccer gear department. He loves soccer, and he can't wait to talk about the sport with customers.

As part of his new job, Ari has a lot of tasks to do. He helps customers and tries to sell them things. He restocks the shelves in his department. He works at the cash registers sometimes.

One of the best parts about the job is the fact that Ari is earning money for the first time. He gets paid $9 an hour, which adds up fast because he's working so much during the summer. How much is Ari making?

Ari works for 30 hours every week. He knows he makes $9 an hour, so he can figure out how much he'll make per week.

$9 x 30 hours = $270 a week

How much money will he make over the whole summer? To find out, you'll need to count up the number of weeks in Ari's summer, which is nine. One of those weeks he'll be on vacation, though. So:

1. $270 x 8 =

Ari actually makes a little more money when he babysits his little sister. His mom pays him $10 an hour to convince him to babysit instead of hang out with his friends. So far, he has babysat these hours over the last month:

Week 1—2 hours
Week 2—4 hours
Week 3—0 hours
Week 4—2 hours

Ari is interested in how many hours, on average, he babysits, so he can guess what he'll make for the rest of the summer. Add up all the hours for the weeks, and then divide by the number of weeks.

2. What is the average number of hours Ari babysits his sister a week?

Now Ari can use the average to figure out how much he'll make babysitting over the summer.

3. How much money will Ari make babysitting? Remember, he has 8 weeks in his summer break.

4. How much money will Ari make in total during the whole summer?

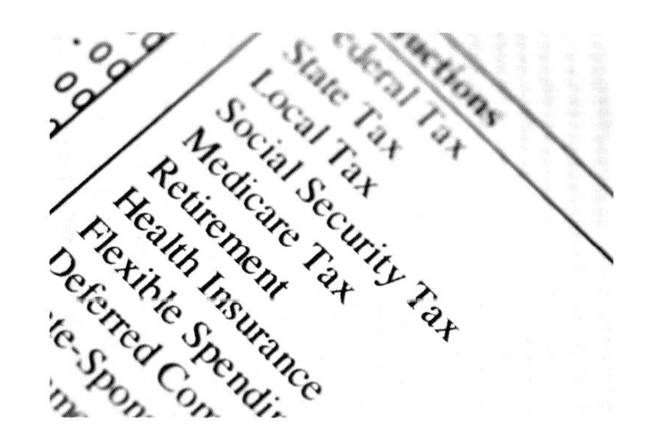

2
WORKING: TAXES

Even though Ari works 30 hours a week at $9 an hour, his first paycheck isn't for $270. He actually only got $206.14. What's going on?

The missing money went to pay taxes. The government collects money from everyone who works. The tax is called an income tax. The government uses taxes to pay for things like the police, schools, roads, bridges, and more. By paying taxes, you're actually helping to pay for all those things you use every day.

Ari can see at the bottom of his pay stub where the tax was taken out. He sees four taxes: federal income tax, state income tax, Social Security tax, and Medicare tax.

Someday Ari wants to own his own sporting goods store. He'll have to pay a lot of attention to taxes then, because business owners have to pay taxes on property (stores and land), on income, and on employees. Right now, though, Ari only has to worry about his own income tax.

Here are the different taxes and tax rates Ari had to pay:

Federal income tax: 10%
State income tax: 4%
Social Security tax: 6.2%
Medicare tax: 1.45%

Percents are parts out of 100, so 40 percent is 40 parts out of 100 parts. The problem with Ari's pay is that it's more than $100. Because he has earned $270, 40% of his pay would be more than just $40.

You can use cross multiplication to figure out how much each tax is. Try it for federal income tax:

$$\frac{10}{100} = \frac{X}{\$270}$$

$\$100 \text{ x } X = 10 \text{ x } \$270$
$\$100 \text{ x } X = 2700$
$X = 2700 \div 100$
$X = \$27$

You can also change percents into decimal numbers by moving the decimal point two spaces to the left. Now 10% becomes .10. Then you can just multiply the number you're starting with by the decimal. For federal income tax, you would get the same answer as before:

$0.10 \text{ x } \$270 = \$27$

Now do the same thing for the other three taxes.

1. How much does Ari pay in state income tax?
2. How much does he pay in Social Security tax?
3. And how much does he pay in Medicare tax?
4. Considering Ari will be making $206.14 every week at the store, after taxes, what will be his new total money earned for the summer?

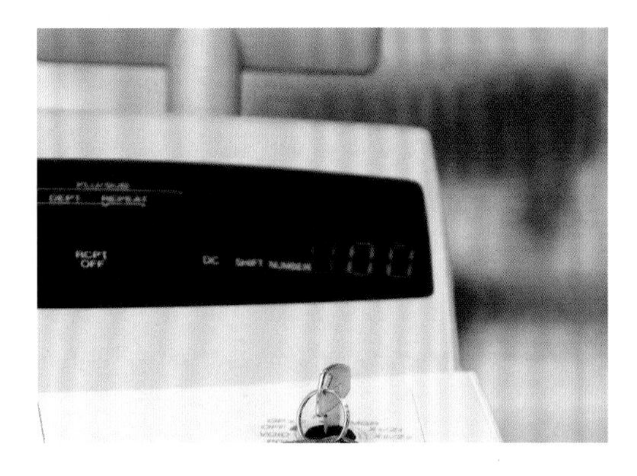

3 WORKING: GIVING CHANGE

Part of Ari's job is to work at the cash register. Normally it's not very hard, and Ari gets to talk to a lot of people, which he likes. He scans the bar codes on the things people are buying, and the cash register adds it all up. Then the customer gives him the money (or a **debit card** or **credit card**) and he gives them change.

One day, the cash registers break! Ari's manager isn't about to close down the store, just because the system the registers use is broken. The cashiers can still open the cash register drawers, take money, and give people change.

Now Ari has to use his math skills to count up the money customers hand him and make sure it's enough. He also has to calculate the change he needs to give back, and he has to do it quickly.

The first customer in Ari's line is buying only one thing, a soccer ball. The ball costs $14.99, and the customer hands over a twenty dollar bill.

1. How much change should Ari give the customer?

The next customer's order is more complicated. Ari has to add up everything she's buying:

Sneakers, $51.50
Tennis racket, $129.00
Tennis balls, $2.59

2. What will the customer have to pay?

The customer doesn't have her debit or credit cards with her. She's going to pay in cash. She hands Ari a fifty dollar bill, six twenties, three dollar bills, and a quarter. He's never seen so much cash all at once!

3. How much did she give Ari? Did she give him enough money?

Ari counts up and tells her she hasn't paid enough. She gives him another twenty dollar bill and asks for her change. Ari does it in his head.

4. How much change will he give the customer?

4 WORKING: CHECKING ACCOUNT

Now that Ari is saving up so much money this summer, he wants to open up a checking account. He already has a savings account where he keeps all the money he's saving for college.

Checking accounts are where you can keep all of the money you are willing to spend. It's safer to keep money in a checking account rather than lying all over your bedroom floor or in your pockets. It's nice to have cash sometimes, but as Ari has discovered many times, you can lose it too.

Ari goes to the bank to set up a checking account. He talks to a bank employee, who tells him about the checking account the bank can offer him. He will have to keep at least $25 in his account at all times, or he will be charged a **fee**. He also has to use his debit card at least twice a month to keep his account active. Otherwise the bank might decide to close the account. He thinks he can do all that, so he signs up and puts $25 in his account. Now all he has to do is keep track of his money, and not spend too much! Check out the next page to figure out how Ari will manage his checking account.

Ari counts up all the money he has in cash at his house. He has a total of $48.83, which he takes to the bank and **deposits** in his checking account. Along with the $25 that's already in there, now he has $73.83.

The next day at work, Ari gets his first paycheck (for $206.14). He goes and deposits half of it in his account. He's saving the other half for college.

1. Now how much does he have in his checking account?

Ari is really excited he has some more spending money. Now he can buy that new mp3 player he's been wanting. It costs $140.

2. Does he have enough money? If so, how much money does he have left?

3. Is the money left more than the $25 minimum he needs in his checking account to avoid a fee?

Ari is out with his friends when they all stop to get ice cream. He hasn't really been paying attention to how much money is left in his checking account, but he thinks he has enough to buy himself some ice cream. Then his friend Luisa asks if he can buy her a sundae and she'll pay him back later. Ari doesn't want to tell her no, so he agrees and spends a total of $12.50 on ice cream and other snacks.

4. Does his checking account still have the minimum amount in it? If not, how low is it?

5 WORKING: BALANCING A CHECKBOOK

Ari is learning about managing money the hard way! After spending too much and paying the $30 fee for not having enough money in his checking account, he's paying a lot more attention to how he spends his money.

When he opened his checking account, Ari got a debit card and also a checkbook. He hasn't used his checkbook yet because he's not sure how. Right now, his uncle happens to be in town, and he sees the checkbook lying on Ari's desk. Uncle Elias tells his nephew he's going to teach him how to write a check and balance a checkbook.

Uncle Elias explains that Ari can check his checking account balance online. He can see what happens when he makes a purchase or a deposit. However, he should also keep track of his purchases and deposits in his checkbook. That way, if the bank or a store makes a mistake, Ari can catch it and make sure he isn't charged too much money. You can see what Uncle Elias teaches Ari on the following pages.

First, Uncle Elias asks Ari to gather together all of his receipts from stores and from making deposits at the bank. Here's what he puts together:

10/27—video game, $35.99
10/29—paycheck, $206.14
11/3—movie ticket, $11
11/7—babysitting money, $30

Now put all of Ari's receipts in the following chart. Purchases go in the withdrawal column, while money added to the account goes in the deposit column.

Date	Description	Withdrawal	Deposit	Balance
10/27	video game	$35.99	—	$74.37
10/29	paycheck	—	$206.14	$280.51

1. How much money does Ari end up with?

2. If Ari wrote a check for $300, would he have enough in his checking account to cover it? If so, how much does he have left? If not, how much more would he need?

6 STARTING A BUSINESS: PROFIT AND COST

Ari's older sister Eva is trying to start a business. She has always wanted to own a business, and now she's finally following her dream. Eva is starting a gardening business, because she loves growing plants.

Eva has already done a lot of planning. She knows she wants to start gardens in people's yards and even in pots on people's porches. She will provide the plants, the garden tools, and the other supplies. She will also teach people how to take care of their gardens.

She isn't sure how much she should charge her customers, though. She doesn't want to charge so much money that people won't be able to afford to hire her. She also wants to make money. Specifically she wants to make a profit, which is the money you make after you account for how much you had to spend to set up your business. How will she figure that out? You'll see how she decides what to charge her customers on the next page.

First, Eva lists all the things she'll have to spend money on to get a job done whenever she is hired. She will be offering two kinds of gardens at different prices— a garden in the yard, and a container garden on a porch, for people who don't have enough yard space or any yard at all.

In-ground garden:
Plants: $150
Compost: $40
Fertilizer: $6

Porch garden:
Plants: $115
Pots: $90
Dirt: $40
Compost: $10
Fertilizer: $6

1. What are her total costs for each type of garden?

She could just charge her customers exactly how much it costs her to build a garden. But then she wouldn't make any money! She also has to account for how much her time is worth. She's putting in effort to start the business and build gardens, which is worth something.

Eva decides her time is worth $20 an hour, because she knows what she's doing and she'll be working hard.

2. If she plans on working an average of 8 hours on in-ground gardens and 3 hours on porch gardens, how much will she make at $20 an hour?

Now you can add the cost of building a garden to Eva's time, to get a good idea of what to charge customers. However, Eva also decides to add on some more money because she thinks customers will pay a little more. She adds $30 to the price of each garden. What should Eva charge customers who buy an in-ground garden?

$$\$196 + \$160 + \$30 = \$386$$

3. What about a porch garden?

Now you can find Eva's profit. You know how much it cost her to buy materials for each garden, and you know how much she is going to charge. The profit for the in-ground garden is the price Eva will charge minus the costs:

$$\$386 - \$196 = \$190$$

4. What profit will she make on the porch garden?

5. Which type of garden should she try to sell more of? Why?

7 STARTING A BUSINESS: OFFICE AREA

To start her garden business, Eva knows she wants to rent a small office. She'll be able to meet with customers, store documents there, and work on the computer from her office. Eventually she can even hire an employee to work in the office, once her business is doing well enough.

She shops around for a few offices. She knows she can only afford a small one, which is fine, because she'll be spending so much of her time outside planting gardens. At most, she can spend $500 a month on rent. Will she be able to find an office she can afford? The next page will help you decide.

The first office Eva looks at is $30 a square foot for the year, and it is 30 feet long and 15 feet wide. Can she afford it?

Here's the formula for square footage:

$$\text{Area (in square feet)} = \text{length} \times \text{width}$$

$$\text{Area} = 30 \times 15$$

1. Area of the office = _______ square feet

Now multiply the square footage by the rental price for the year and divide it by twelve months, to find the monthly rental cost.

2. Is it within Eva's budget?

Eva looks at a second apartment that is $18 a square foot for the year, and is 330 square feet.

3. Can she afford this one? Why or why not?

Eva needs space for two desks, three filing cabinets, and a table and chairs in her office. The desks are 3.5 feet x 1.5 feet. The filing cabinets are 1 foot x 1 foot. And the table and chairs take up 5 feet by 4 feet of space. And she wants to still be able to walk around, so she should have at least 100 square feet of extra room. Will this office have enough room for all of her stuff?

Area of desks = 2 x (3.5 x 1.5)

4. Area of desks =

5. Area of filing cabinets =

6. Area of table and chairs =

7. Now add up all the areas. Don't forget to take the walking space into account.

8. Does the office have enough room? If so, how much extra space does it have? If not, how much more space would it need.

8
STARTING A BUSINESS: LOANS

Eva doesn't quite have enough money to start her business. For example, she has to pay for wheelbarrows, hoses, and other garden tools. She also has to be able to buy the plants and pay the rent on her office for her first couple months in business. She thinks once she gets her business going, she'll be able to make plenty of money. First, though, she has to start making money. And businesses need money to make money.

Fortunately, Eva can get a loan from the bank. Loans are borrowed money. When she starts making money with the garden business, Eva will be able to pay the bank back.

Eva goes to the bank to ask about loans. She talks to a bank employee named Miguel about how much she should take out. Miguel takes her through some calculations. Help her figure it out on the next page.

Eva first has to come up with how much money she needs, so she can figure out how much in loans she should get.

Here's a list of all the expenses Eva will have to start her business:

first month's office rent: $495
wheelbarrow: $45.50
hoses: $19.99
pitchfork: $32.99
materials for first few gardens: $600
office supplies: $35
2 desks: $80 each
2 desk chairs: $48.75 each
3 filing cabinets: $17.50 each
table and chairs: $160.00
printer: $200.00
advertising: $50

gas for her truck: $60
business licenses: $150

1. How much will Eva need all together?

However, Eva won't need to borrow all that money. She has been saving up to start her business for a while. She has already spent some of it, but she has some left to pay for what she needs now that she's almost ready to open. In total, she has $600 saved.

The difference between the amount Eva needs to start her business and the money she has saved is how much she should take out in loans.

2. How much money should Eva get in loans?

9 STARTING A BUSINESS: INTEREST

While Eva is at the bank, Miguel also tells her about loan interest. Interest is the extra fee you have to pay just to take out a loan. You don't pay interest right away—you add it on to your loan payments when you start paying them back.

A high interest rate means Eva will have to pay a lot of extra money for her loan. A lower interest rate means she will have to pay less.

Miguel also explains that the loans the bank offers come with compound interest. Eva will be taking out $1560 in loans. Miguel tells her the bank's interest rate is 7.5%, and that she will have two years to pay off her loan. Her first interest payment will just be based on that original $1560 in loans. The second year, she will have to pay more interest. She will have to pay interest on the original loan, plus the interest she owes from the first year. The bank sees the interest as a second kind of mini-loan, so it charges interest on that. That's what "compound interest" means.

You can see interest can add up quickly! Check out the next page to find out how much Eva will have to pay in interest to take out a loan.

The math equation you use to calculate how much money you will have after interest is:

$$\text{total amount} = P\,(1 + r/n)^{nt}$$

That may look complicated, but it's not too bad once you know what all those letters mean.

P = how much money you are taking out, called the principal
r = the rate of interest, which the bank tells you; r is always a decimal.
n = the number of times interest is calculated over time. It might be calculated once a month or four times a year, or once a year.
t = the number of years you're looking at

The little nt is an exponent, which tells you how many times you will multiply a number by itself. 23 would be 2 x 2 x 2, which equals 8. Your calculator can do exponents for you.

Eva's information is:

P = \$1560
r = 0.075 (from 7.5%)
n = 12, because interest will be calculated monthly
t = 2 years

Find the total amount of money Eva will owe, including interest, after two years:

1. Total amount = $\$1560\ (1 + .075/12)^{12 \times 2}$

   Total amount =

2. How much of that total is interest?

10 RUNNING A BUSINESS: SALES

Now Eva has her business up and running. She has started advertising, set up her office, and is planting her first gardens for customers. Eva wants to keep careful records of her business, so she knows exactly what's going on.

She especially is interested in her sales, which is the money she's making from customers. The more customers she has, the more jobs she can do and the more money she can make. The first month of her business is a little slow, but by the second, she has a lot of customers.

Eva can calculate how much profit she's making, whether she's advertising well, and if she can make loan payments to the bank. Take a look at her records and see how much she's making in sales.

The first month, Eva has just three customers. She builds two in-ground gardens, and one porch garden.

1. How much does she make the first month? Look back to page 19 to see how much each garden costs.

At the beginning, Eva calculated how much she would charge each customer based on how much each garden cost to build. However, she forgot to include her monthly loan payments. She has to pay $75 each month, in order to pay off her whole loan in two years. After the first month, she'll also have to pay rent and gas money.

She adds up the cost of building each of the gardens, and then adds it to her first month's loan payment. If it's less than the money she made this month, the difference is the profit.

2. Did Eva make any profit? If so, how much? If not, how much more did she need?

Her second month, Eva spends more time advertising. She ends up building five in-ground gardens and four porch gardens. But this month, she also had to pay for office rent and $60 worth of gas, plus the loan payment and the cost of garden materials.

3. Did Eva make a bigger profit this month than last, or did she make less?

11 RUNNING A BUSINESS: ORDERING SUPPLIES

Every time she gets hired, Eva has to order supplies to build a new garden. She works with her customers to figure out just what they want and what will work for their space. Some customers want flower gardens in their yard, some want bushes right outside the door, and some want vegetables in pots on the porch.

Eva is still figuring out what the best suppliers are. She likes to buy flowers from one supplier, vegetable seedlings from another, pots from another, and dirt from yet another. Sometimes she gets really confused about what is coming from where, but she knows it's important to keep track of it all.

To make sure she's paying the right amount of money, Eva always checks her order forms when her materials arrive. She has found mistakes a few times, so she knows it's worth taking the time to check.

Usually Eva fills out an order form that looks like this:

Date: 5/16
Order Number: 12345678

Description	*Quantity*	*Price*	*Total*
Squash seedlings	4	$3.99 ea.	$15.96
Tomato seedlings	5	$6.99 ea.	$34.95
Eggplant seedlings	3	$5.50 ea.	$16.50
Pea seedlings	10	$2.59 ea	$25.90
Radish seeds	1	$3 per package	$3
Lettuce seeds	1	$3.50 per package	$3.50
Total cost:			$99.81
Tax (8%)			$7.98

Shipping and handling	$6.54
Payment due	$114.33

What she actually gets is:

4 squash seedlings
4 tomato seedlings
0 eggplant seedlings
10 pea seedlings
1 package radish seeds
0 packages lettuce seeds

The company really messed up Eva's order!

1. What are the differences between the order form and what she actually receives by truck?

2. What is the price difference between what she ordered and what she got?

12 RUNNING A BUSINESS: DISCOUNTS

Eva is talking to her brother Ari about her business. She's been doing pretty well, but she wants a few more customers. She's been doing a lot of advertising, and she's not sure what else she can do to convince more people to hire her.

Ari has a great idea—he suggests she offer customers a discount. That way, the customers who thought Eva's prices were a little too high will decide they can afford a garden after all.

Eva agrees, and gets to work right away. On the next page, figure out how much of a discount she should advertise.

Eva will offer a certain percentage off her gardens. First she considers offering a 25% off coupon in the paper.

How much would 25% off be for a customer who ordered an in-ground garden?

You can calculate 25% off a couple different ways. First, you could think of the percentage in terms of fractions: 25% of something is the same as ¼ of something. So, for an in-ground garden, that costs $386:

$$\$386 \div 4 = \$96.50$$

Next, subtract the money customers will save from the total price of the garden:

$$\$386 - \$96.50 = \$289.50$$

You can also **convert** percentages into decimals. This means 25% would be .25. Then multiply the original cost by .25, and subtract that from the original cost. You'll get the same answer as before.

$$.25 \times \$386 = \$96.50$$
$$\$386 - \$96.50 = \$289.50$$

Now try it for the porch garden.

1. How much will customers pay if they get 25% off?

Eva only wants to offer a 25% off discount if she still makes a profit. Then she will still be making money, and also getting new customers who will tell their friends what a wonderful business she runs.

2. Will Eva make a profit off the in-ground garden if she offers a 25% discount? If so, how much of a profit will she be making after the discount? Turn back to page 19 for the numbers you need. Assume you don't need to take her monthly loan payment into account.

3. Will she make a profit from the porch garden? If so, how much?

4. Based on your answers, do you think Eva should focus on promoting 25% off on in-ground gardens or porch gardens?

13 RUNNING A BUSINESS: COMMISSION

Eva thinks her brother Ari is a good businessperson. She needs a little help now that her business is picking up, and she has a lot of customers. She asks Ari if he would want to work for her on commission.

She explains that commission isn't exactly like being hired. Instead, Ari will try to sell gardens to customers. Every time he sells a garden, he will earn a little money. The money he will earn is the incentive for him to work hard at selling gardens. The harder he works, the more money he earns.

Ari thinks it over and decides he wants some more money. He isn't working as many hours at the sporting goods store because he's back at school. By working on commission for his sister, he can choose what hours he works and still save up for college and get some spending money.

Employers can pay workers the same amount every time they make a sale. Or they can pay the worker a percentage of the sale, like 5% or 15%. The next page will show you how Eva figures out how much to pay Ari.

Eva will pay Ari a certain amount of money each time he convinces a customer to buy a garden. How much should she pay him for each garden?

She decides to pay Ari a percentage of her sales. She only sells two products, so it won't be hard to figure out how much she needs to pay him.

First, she tries out a commission rate of 1%.

What would 1% of an in-ground garden sale be? Use decimals.

1. .01 x $386 =

2. Do you think Ari would be wasting his time trying to sell gardens for that much commission? Does he have other choices that would allow him to make more money,

considering it takes him two hours to make a sale?

Eva sees paying a 1% commission is too little.

Now calculate a 10% commission on an in-ground garden:

3. .10 x $386 =

4. What about a 10% commission on a porch garden?

5. How much would Ari make on an in-ground garden that was discounted by 25%?

Ari decides a 10% commission sounds good to him, and he takes the job.

14 RUNNING A BUSINESS: EMPLOYEES

After just a few weeks, Ari brings in seven new customers to the garden business. Eva barely has time to keep up with all the new work, and she's starting to fall behind. Then she gets an idea—maybe she should hire her brother to work for her part time. He could do sales, and he could also help build the gardens.

Hiring an employee can cost a business a lot of money. On the other hand, her business can't grow if she can't do all the work. Ari has already made almost three hundred dollars working on commission, so she knows she can afford at least a few hours. She does a few calculations to see how many hours a week she can afford to hire her brother.

Eva knows she has to offer her brother at least the same amount of money per hour he is making at the sporting goods store. Otherwise, he has no reason to leave that job. So, she has to offer him more than $9 an hour.

1. How much would she be paying him a week if she paid him $11 an hour and hired him for 10 hours a week?

2. How much more would Ari make working for Eva than if he worked 10 hours at the sporting goods store?

Now Eva has to decide if she can afford to hire her brother for 10 hours a week, at $11 an hour.

She figures Ari will be working about four weeks a month, so she will pay him $440 a week. Eva also expects she'll be able to build more gardens with an employee's help. She **estimates** her business can build six more gardens a month.

Find the range of profits she could have. The range is the lowest possible profit to the highest possible profit. First, calculate what her profit would be if she builds six porch gardens, which make less profit. This is the bottom limit of the range. Turn to page 19 to find the profit numbers.

3. 6 x $90 =

Now do the math to find out if the six gardens were in-ground gardens, which make a higher profit. This is the upper limit of the range.

4. What is the upper limit of the range?

5. What is the range of profits she could make by hiring Ari?

6. Do you think hiring Ari is worthwhile? Why or why not?

15 PUTTING IT ALL TOGETHER

Eva ends up hiring Ari, and her business picks up even more. Eva's business is a success! She has come a long way from her first days starting the garden business. And Ari has come a long way from his first days working as an employee.

See if you can remember what Ari and Eva have learned about working and running a business.

1. If Ari were to get a raise to $13 an hour, how much would he make working 10 hours a week?

2. With the same tax rates he had before, how much would Ari's weekly paycheck actually have?

3. Ari puts 35% of his weekly paycheck into his checking account. How much does he have after 4 weeks, if he doesn't spend any of it?

4. Will Ari have enough money in his account if he deposits a $50 check his grandma gave him for his birthday—and then he buys a TV that costs $200?

5. The flower pots Eva usually buys for her customers suddenly go up in price. They are now $5 more each, and she usually buys 5 of them. How much will her profits be reduced by, if she doesn't raise the price of her gardens?

6. Eva's in-ground gardens are usually 6 feet by 8 feet. What is the area of the gardens in square feet?

7. How much would Eva have to pay each month on her loan if she only had a year to pay it off, and everything else was the same?

8. Eva is offering a new garden sale: every in-ground garden is 15% off. How much would one in-ground garden be?

Answers

1.

1. $2160

2. 2 hours
3. $10 x 2 hours x 8 weeks = $160
4. $2160 + $160 = $2320

2.

1. $10.80
2. $16.74
3. $3.92
4. ($206.14 x 8)+160 = $1809.12

3.

1. $20 – $14.99 = $5.01
2. $51.50 + $129.00 +$2.59 = $183.09
3. No. She gave him $173.25.
4. $193.25 – $183.09 = $10.16

4.

1. $73.83 + ($206.14/2) = $176.90
2. Yes, he'll have $36.90 left over.
3. Yes.
4. No, he has $24.40, which is $0.60 less than he needs to have.

5.

1. $299.51
2. He doesn't have enough. He needs $25.49 more, including the money to the minimum balance that needs to be in his account.

Date	Description	Withdrawal	Deposit	Balance
10/27	video game	$35.99	—	$74.37
10/29	paycheck	—	$206.14	$280.51
11/3	movie ticket	$11	—	$269.51
11/7	babysitting money	—	$30	$299.51

6.

1. In-ground garden: \$196, Porch garden: \$261.
2. In-ground garden- \$20 x 8 = \$160, Porch garden- \$20 x 3 = \$60
3. \$261 + \$60 + \$30 = \$351
4. \$351 – \$261 = \$90
5. The in-ground garden, because she can make more profits.

7.

1. 450
2. No, it will cost \$1125 a month.
3. Yes, it is \$495 a month.
4. 10.5 square feet
5. 3 square feet
6. 20 square feet
7. Yes, it has plenty of space (10.5 + 3 + 20 + 100 = 133.5 square feet).
8. It has 196.5 extra square feet!

8.

1. \$2158.48
2. \$2158.48 – \$600 = \$1558.48

9.

1. \$1811.62
2. \$1811.62 – \$1560 = \$251.62

10.

1. (2 x \$386) +\$351 = \$1123
2. Yes, she made a profit of \$395 her first month.
3. (5 x \$386) + (4 x \$351) = \$3334
   \$3334 – (\$75 + (5 x \$196) + (4 x \$261) +\$495 + \$60) = \$680

She made $285 more in profit this month.

11.

1. 1 tomato seedling, 3 eggplant seedlings, and a package of lettuce seeds are missing.
2. $6.99 + $16.50 + $3.50 = $26.99

12.

1. $263.25
2. Yes, she will still make a $93.50 profit.
3. Yes, but only $2.25.
4. She should focus on discounting in-ground gardens, because she will make more money even with the discount.

13.

1. $3.86
2. He could be working at the sporting goods store or babysitting and making more money per hour.
3. $38.60
4. .10 x $351 = $35.10
5. .10 x $289.50 = $28.95

14.

1. $11 x 10 = $110
2. $20 more a week.
3. $540
4. $1176
5. $540 – $440 = $100 (lower limit), $1176 –$440 = $736 (upper limit)
6. Yes, because the business will make a profit no matter what.

15.

1. \$130
2. \$13 + \$5.20 + \$8.06 + \$1.89 = \$28.15, \$130 – \$28.15 = \$101.85
3. 4 x (.35 x \$101.85) = \$142.59
4. No, he doesn't have enough to buy the TV.
5. \$5 x 5 = \$25.
6. 6 feet x 8 feet = 48 square feet
7. Total amount = \$1560 $(1 + .075/12)^{12 \times 1}$

   Total amount = \$1681.11

   \$1681.11/12 months = \$140.09
8. .15 x \$386 = \$57.90, \$386 –\$57.90 = \$328.10

本书由中国科学院数学与系统科学研究院资助出版

数学 24/7

银行中的数学

〔美〕海伦·汤普森　著

孙云志　译

科学出版社

北　京

图字：01-2015-5622号

内 容 简 介

银行中的数学是“数学生活”系列之一，内容涉及如何使用零用钱和红包、如何兼职赚钱、如何在银行开户及计算利率、如何使用ATM机及手续费、如何购买股票和计算利润，以及使用行用卡和货币兑换等知识，让青少年在学校学到的数学知识应用到与银行有关的多个方面，让青少年进一步了解数学在日常生活中是如何运用的。

本书适合作为中小学生的课外辅导书，也可作为中小学生的兴趣读物。

图书在版编目（CIP）数据

银行中的数学/（美）海伦·汤普森（Helen Thompson）著;孙云志译.—北京:科学出版社,2018.5
（数学生活）
书名原文：Banking Math
ISBN 978-7-03-056745-1

Ⅰ.①银… Ⅱ.①海… ②孙… Ⅲ.①数学-青少年读物 Ⅳ.①01-49

中国版本图书馆CIP数据核字（2018）第046668号

责任编辑:胡庆家 / 责任校对:邹慧卿
责任印制:肖　兴 / 封面设计:陈　敬

科 学 出 版 社 出版
北京东黄城根北街16号
邮政编码：100717
http://www.sciencep.com

北京汇瑞嘉合文化发展有限公司 印刷
科学出版社发行　各地新华书店经销
*
2018年5月第　一　版　　开本:889×1194 1/16
2018年5月第一次印刷　　印张:4 1/2
字数:70 000

定价：98.00元（含2册）
（如有印装质量问题，我社负责调换）

引 言

你会如何定义数学？它也许不是你想象的那样简单。我们都知道数学和数字有关。我们常常认为它是科学，尤其是自然科学、工程和医药学的一部分，甚至是基础部分。谈及数学，大多数人会想到方程和黑板、公式和课本。

但其实数学远不止这些。例如，在公元前5世纪，古希腊雕刻家波留克列特斯曾经用数学雕刻出了“完美”的人体像。又例如，还记得列昂纳多·达·芬奇吗？他曾使用有着赏心悦目的尺寸的几何矩形——他称之为“黄金矩形”，创作出了著名的画作——蒙娜丽莎。

数学和艺术？是的！数学对包括医药和美术在内的诸多学科都至关重要。计数、计算、测量、对图形和物理运动的研究，这些都被融入到音乐与游戏、科学与建筑之中。事实上，作为一种描述我们周围世界的方式，数学形成于日常生活的需要。数学给我们提供了一种去理解真实世界的方法——继而用切实可行的途径来控制世界。

例如，当两个人合作建造一样东西时，他们肯定需要一种语言来讨论将要使用的材料和要建造的对象。但如果他们建造的过程中没有用到一个标尺，也不用任何方式告诉对方尺寸，甚至他们不能互相交流，那他们建造出来的东西会是什么样的呢？

事实上，即便没有察觉到，但我们确实每天都在使用数学。当我们购物、运动、查看时间、外出旅行、出差办事，甚至烹饪时都用到了数学。无论有没有意识到，我们在数不清的日常活动中用着数学。数学几乎每时每刻都在发生。

很多人都觉得自己讨厌数学。在我们的想象中，数学就是枯燥乏味的老教授做着无穷无尽的计算。我们会认为数学和实际生活没有关系；离开了数学课堂，在真实世界里我们再不用考虑与数学有关的事情了。

然而事实却是数学使我们生活各方面变得更好。不懂得基本的数学应用的人会遇到很多问题。例如，美联储发现，那些破产的人的负债是他们所得收入的1.5倍左右——换句话说，假设他们年收入是24000美元，那么平均负债是36000美元。懂得基本的减法，会使他们提前意识到风险从而避免破产。

作为一个成年人，无论你的职业是什么，都会或多或少地依赖于你的数学计算能力。没有数学技巧，你就无法成为科学家、护士、工程师或者计算机专家，就无法得到商学院学位，就无法成为一名服务生、一位建造师或收银员。

体育运动也需要数学。从得分到战术，都需要你理解数学——所以无论你是想在电视上看一场足球比赛，还是想在赛场上成为一流的运动员，数学技巧都会给你带来更好的体验。

还有计算机的使用。从农庄到工厂、从餐馆到理发店，如今所有的商家都至少拥有一台电脑。千兆字节、数据、电子表格、程序设计，这些都要求你对数学有一定的理解能力。当然，电脑会提供很多自动运算的数学函数，但你还得知道如何使用这些函数，你得理解电脑运行结果的含义。

这类数学是一种技能，但我们总是在需要做快速计算时才会意识到自己需要这种技能。于是，有时我们会抓耳挠腮，不知道如何将学校里学的数学应用在实际生活中。这套丛书将助你一马当先，让你提前练习数学在各种生活情境里的运用。这套丛书将会带你入门——但如果想掌握更多，你必须专心上数学课，认真完成作业，除此之外再无捷径。

但是，付出的这些努力会在之后的生活里——几乎每时每刻（24/7）——让你受益匪浅！

目　录

Contents

1
如何用好零花钱

杰米过去每周能得到2美元的零花钱。她花一半，存一半。事实上，她有两个存钱罐，一个用来放零花钱，一个用来存钱。最近，杰米的父母给她增加了零花钱。他们觉得杰米一直对零花钱的使用很合理，现在她也长大了，所以她每周可以得到更多的零花钱。杰米的父母决定每周给杰米10美元。

杰米喜欢购物，因此她觉得非常开心。她也希望可以存下一部分零花钱，这样将来就可以买一些更贵的东西。另外，杰米是个乐于助人的孩子，她还希望把存下的一部分零花钱捐献给慈善机构。

杰米知道10美元比2美元多了很多，但是不确定这些加起来一共是多少。一年下来，她能多得到多少零花钱？

首先她需要知道，每周2美元的话，她可以得到多少。一年有52个星期，那么

$$52\text{周} \times 2\text{美元/周} = 104\text{美元}$$

现在她每周可以得到10美元：

$$52\text{周} \times 10\text{美元/周} = 520\text{美元}$$

1. 杰米知道她得到了很多钱，但是到底比原来多了多少呢？请你来计算一下。

杰米决定将这笔钱的四分之一存起来。四分之一就好比是把一个物体分成四份。她把自己的零花钱等分成了四份，其中一份放进储蓄罐里。这只需要把她每周的零花钱除以4就可以了。

2. 她每周可以存下多少钱呢？

现在她手上还剩四分之三的零花钱。她希望每周捐1美元给“食品橱柜”，去帮助那些有需要的人们。

3. 除去存下的钱和捐款，现在她每周还剩下多少钱用来花销呢？

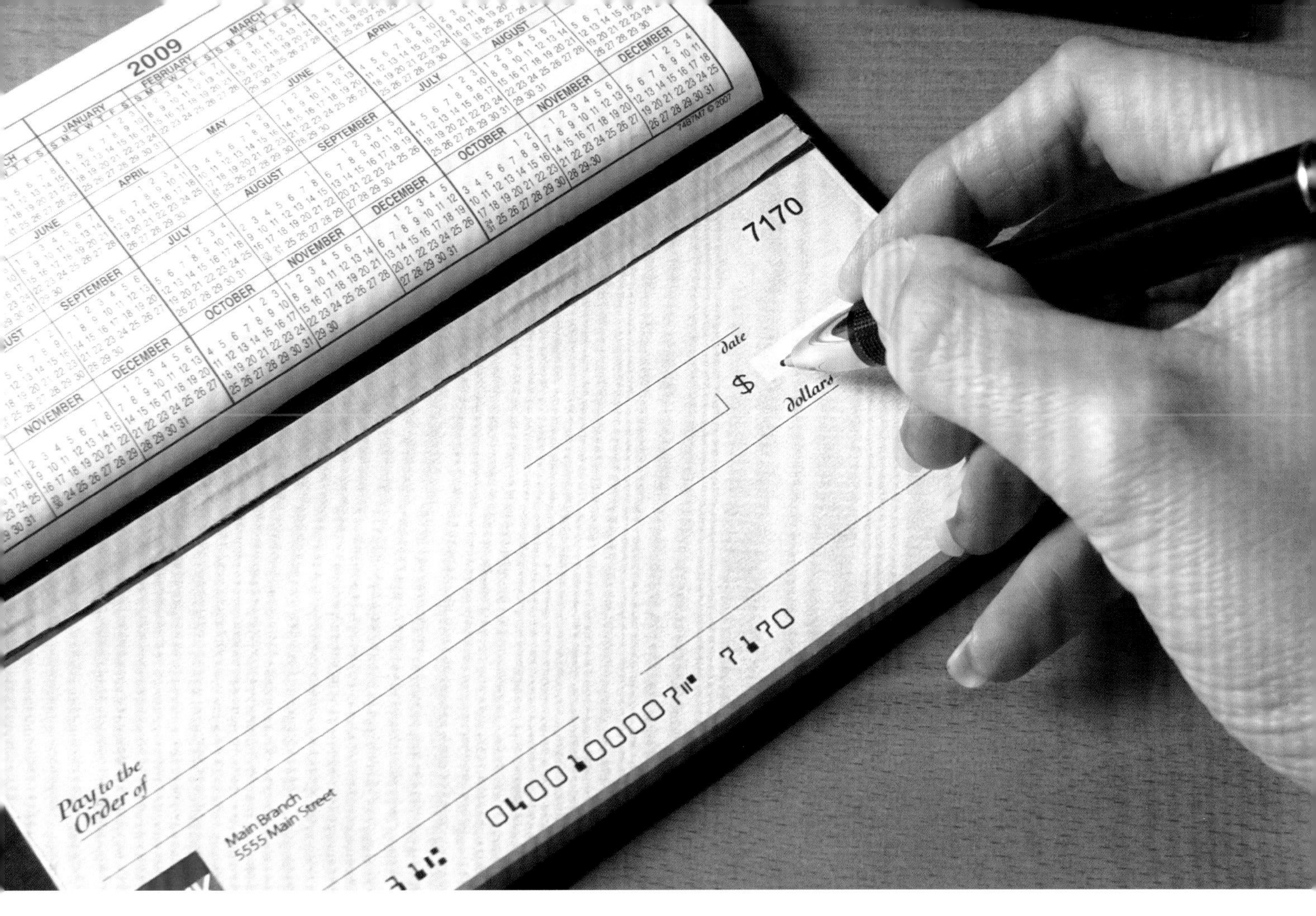

2

如何使用红包

还有一周就是杰米的生日了。她收到了家人和朋友们寄来的生日贺卡。奶奶寄来的卡片里还夹着一张50美元的支票。爸爸妈妈也提前准备了生日礼物。他们知道杰米从不乱花钱，所以给了她一张40美元的支票，不过他们告诉杰米必须把这笔钱存起来。杰米的叔叔也给了她一张35美元的支票，让她自己买件喜欢的东西作为生日礼物。

于是，杰米一下子得到了一大笔钱，比过去任何时候都要多。她去银行兑换了所有支票，一共得到了125美元现金。现在她开始考虑该怎么使用这笔钱了。

杰米共计收到了三张支票，金额分别是50美元、40美元和35美元。其中两张需要按照事先约定去使用：40美元的支票是要存起来不能马上花掉的，而35美元的支票是用来给自己买件生日礼物的。杰米非常想要买一个新的视频游戏，需要50美元。但是35美元还不够，她还差多少钱呢？

50美元-35美元=？

通过简单的计算，可以知道她还需要15美元。她不能使用爸爸妈妈给的支票，不过可以使用奶奶给的那张。

1. 买了视频游戏之后，她还剩下多少钱？我们可以通过以下步骤来计算：

叔叔的支票余下的部分+奶奶的支票余下的部分+父母的支票余下的部分=？

(35美元-35美元)+(50美元-15美元)+40美元=

记住，计算时需要先计算括号内的公式。

2. 或者，我们从大处着眼：

总收入125美元-购买视频游戏应支出的50美元=

可以知道，杰米一共还剩余75美元可以存起来。加上之前已经存了四个星期的零用钱。

3. 她一共存下了多少钱？（在第1节中我们计算过杰米每周存钱的金额）

3
如何挣外快

杰米因为最近的收入而非常兴奋。她告诉了好朋友卡里姆。卡里姆说他也希望有些收入，这样他就可以买一辆心仪已久的自行车了。经过讨论，他们决定一起去找份挣钱的事情来做。

杰米想找份在家附近的活儿。她的爸爸刚找了一份全职的工作，所以没时间打扫卫生和洗衣服。杰米问爸爸是否可以做一些杂务来换取报酬。经过一番考虑后，爸爸决定让杰米每周清除屋外草坪的杂草，每周洗一次衣服，隔一阵子帮助爸爸妈妈打扫一次房间。杰米每工作一小时可以得到5美元作为报酬。

与此同时，卡里姆决定帮别人遛狗来挣些外快。他认识所有邻居，很多人都养了狗。几个月前，他曾答应一个邻居帮她遛狗，并得到了一些报酬。卡里姆意识到他可以主动出击去寻找那些需要遛狗的邻居们。他问了一圈儿，有三个邻居答应了让他每周帮着遛一次狗。第一个邻居每次给他5美元，第二个每次给他6美元，第三个邻居有两只大狗，所以每次给他12美元。

于是杰米和卡里姆都开始挣钱了。他们一共可以挣多少呢？

杰米每周做三个小时的零工。

1. 她可以挣多少钱？如果她希望挣40美元，每周需要工作几个小时？

卡里姆这边报酬各不相同。他收入的数额取决于不同的邻居给的报酬。请按照第一个表格中的例子，填写后面两个表格，计算卡里姆为不同邻居工作获得的收入情况：

工作时间(邻居1)/小时	报酬/(5美元/小时)
1	5美元
2	10美元
3	15美元

工作时间(邻居2)/小时	报酬/(6美元/小时)
1	
2	
3	

工作时间(邻居3)/小时	报酬/(12美元/小时)
1	
2	
3	

2. 卡里姆需要为第一个邻居工作几个小时才能得到和杰米一样的收入？

3. 他需要为第三个邻居工作几个小时才能比杰米挣得更多？

4. 他需要为第二个邻居工作几个小时才能每周挣100美元？

5. 在一个学年里，他有没有那么多时间来做这份工作呢？

4
如何在银行开户

因为善于理财，杰米现在有了一大笔钱，她的储蓄罐已经放不下了。对她来说，现在有必要去银行开一个存款账户了。

爸爸告诉她把钱放在银行存款账户里是个不错的做法。虽然放在账户里的钱看不见摸不着，但是却非常安全。当她不需要用钱的时候，也不需要去管它们。

杰米和爸爸一起去了银行，咨询工作人员适合杰米办理的账户类型。银行方面提供了一些选择，其中有些账户需要按月收费，另一些则要求账户中保持一定金额才可以免费。如果杰米不经常使用她的账户，就需要支付一定的费用。

杰米比较了各种不同类型的银行账户来选择最适合自己的。首先，她研究了需要按月付费的账户，费用是每月10美元，一年后变为每月8美元。

我们来计算一下她两年期间需要付给银行多少管理费？

1. (10美元/月 × 12月)+(8美元/月 × 12月)=

2. 和杰米存下的钱相比，这些手续费是多还是少呢？如果多的话，多了多少呢？

另一种可以选择的存款账户需保持至少50美元的存款。如果账户中少于50美元，则需要支付30美元的管理费。现在，她已经有了140美元，包括零花钱、生日礼金，以及做零工的报酬。她计划继续做些零工，每周3个小时，每个小时5美元。同时，她计划下个月(4个星期内)给自己买一台价值100美元的电视机。从现在算起的四个星期之后，结余的存款是否足以保证她免交管理费呢？

计算一下，她做四个星期的零工可以收入多少？

3. 3小时/周 × 5美元/小时 × 4周=

4. 现在再算上她已经有的存款。她一共有多少钱？

5. 买了电视机之后，她还剩下多少存款？

6. 和存款账户最低存款50美元相比，是多还是少呢？多了多少呢？

7. 根据你的答案，你认为杰米应该选择哪种账户呢？为什么？

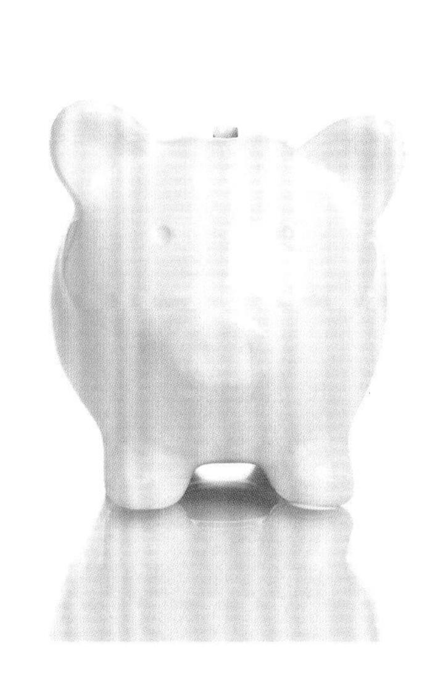
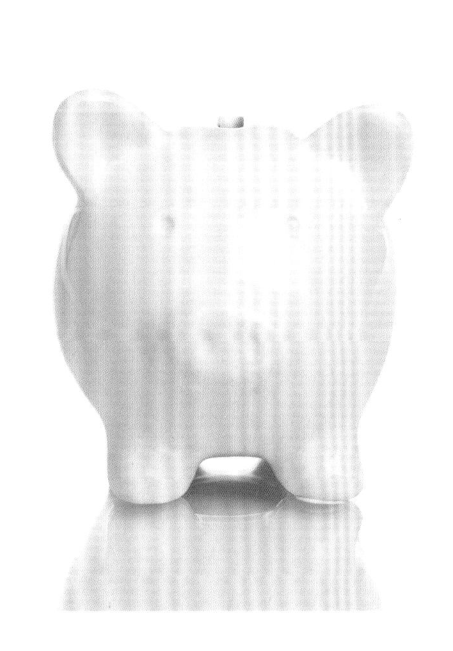
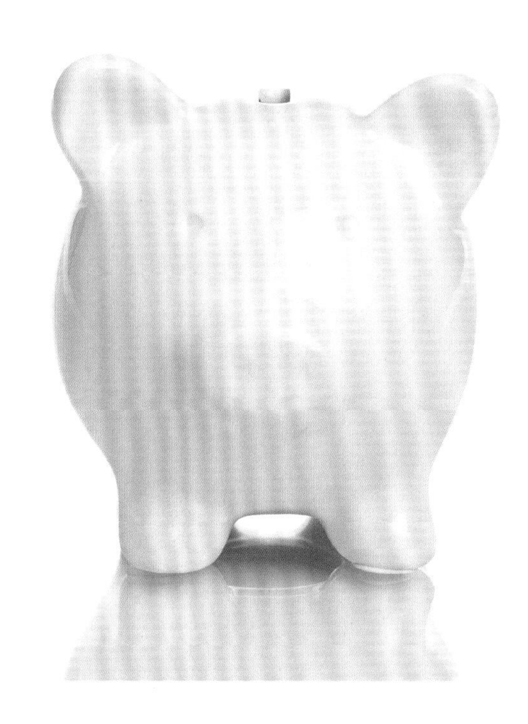

5
利　　率

杰米在银行办理了一个新的存款账户。银行的工作人员告诉她，她的存款放在账户里会定期收到一些利息。利息是银行付给存款客户本金以外的一笔钱。存的钱越多，就可以获得越多的利息，并且利息部分还会得到新的利息。杰米现在已经迫不及待想把她攒的钱存进账户了。

家里存钱罐里的钱是不会有利息的。回到家以后，她一直在问爸爸关于利息的问题。她想知道按现有的存款，可以得到多少利息。爸爸告诉她，不会太多，但是总会有一些。下面我们来看看如何计算利息。

我们用下面的公式来计算利息：

$$总金额=P(1+r/n)^{nt}$$

看起来有点儿复杂，但是如果了解每个字母代表的含义，其实也不是很困难：

- P 表示存款账户中的总金额
- r 表示银行存款利率
- n表示存款时间内计算利息的次数。可以按月、季度或者年度来计算
- t 表示存款年限

杰米的利率公式如下：

P = 140美元，存款账户中的金额
r = 0.3%，或者0.003，也可以说是百分之一的三分之一
n =12，因为她的利率是按月计算的，一年有12个月
t =1年

现在我们来计算杰米账户中的140美元一年下来可以得到多少利息。公式末尾右上部字号较小的数字是指数，也可以写成^nt。可以用计算器来计算指数。

$$总金额=140美元(1+0.003/12)^{12\times1}$$
$$总金额=140.42美元$$

杰米得到了多少利息呢？

1. 140.42美元 - 140.00美元 =

2. 如果杰米将140美元存入银行三年，她将会有多少钱？

3. 如果利率是1%(0.01)，一年后她将会有多少钱？

4. 如果存500美元的话，她将会有多少钱？

6
如何使用支票账户

几个月后，杰米存了一大笔钱。为了方便买东西，她也放了一些现金在家里。

现金保管起来并不容易。杰米曾经弄丢了一张20美元钞票，这是她做零工挣来的。她找遍了家里的每一个角落，最终还是没有找到。她也经常能在屋子的角落里发现一些现金，床的下面、背包里面，甚至是壁橱的一角。

杰米在学校和好朋友玛雅聊天的时候提起了她的这个问题，玛雅告诉杰米她也有同样的问题，她去银行开了一个支票账户，支票账户同时会配发一张借记卡。玛雅给杰米解释说，借记卡的功能类似于现金，可以在商场里使用借记卡直接花费支票账户里的钱。

玛雅的支票账户和存款账户是分开的，虽然都是在同一家银行办理的。她通常不会从存款账户里取钱。她存钱是为了将来上大学用的。然而，她用钱总是从支票账户中支取，并且有了收入就会存更多进去。她很少用钞票和硬币，需要付钱买东西的时候，她通常都是使用银行卡。

杰米和爸爸又去银行办理了一个支票账户。杰米能否记录下支票账户中的资金流水呢？使用银行卡来代替现金支付，会不会让她花掉更多的钱呢？

经过了几个月的储蓄，杰米现在有了很多零用钱。尽管买了一些东西，比如视频游戏，她现在一共还有350美元。现在她有了一张崭新的借记卡，想找个地方试一下。她去商场买了几件衣服，付钱的时候没有注意标价。下面是她买的东西：

牛仔裤，29.99美元
2件T恤衫，每件15.99美元
鞋子，34.99美元

1. 她一共花了多少钱？不要忘了T恤衫的价格需要乘以2。

杰米的支票账户中还剩下多少钱？她存了350美元，花掉了96.96美元，所以还剩下253.04美元。

杰米在商场花了不少钱。你可以算算她花掉的钱占了多大的百分比。百分比是一种表达比例的方式。例如，如果杰米有100美元，花掉了40美元，那么她花掉了40%。然而，杰米的钱比100块要多，所以我们希望知道她花费的百分比。

你需要通过叉乘法来计算：

$$\frac{96.96\text{美元}}{350\text{美元}} = \frac{X}{100}$$

求解X。

2.

$$96.96\text{美元} \times 100 = 350\text{美元} \times X$$
$$X=(96.96\text{美元} \times 100)/350\text{美元}$$
$$X=\_\_\_\_\_\%$$

3. 杰米花掉的钱的百分比大于她的支票账户中金额的一半吗？（一半即50%）

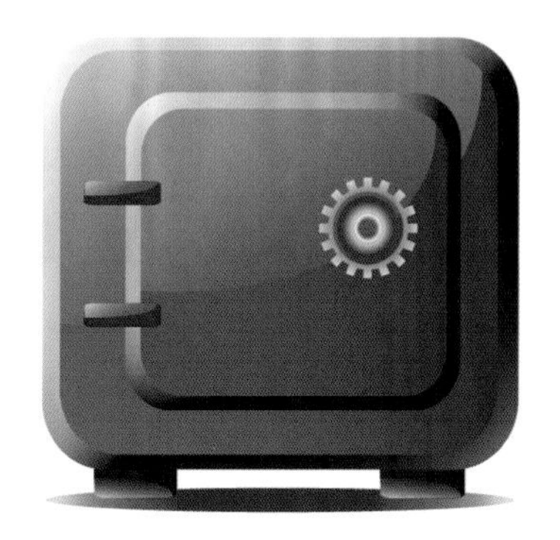

7

收支平衡

杰米办理支票账户的同时，得到了一个崭新的支票本。虽然她完全不知道该怎么用，但是非常有兴趣去了解一下。表哥布兰顿来看她，于是她问了他关于支票的事。

布兰顿告诉杰米他经常使用支票。他租了一间公寓，每次付房租的时候，他就会寄一张支票给房东。他每个月还会用支票来支付他的电费和暖气费。有时，因为忘了带银行卡和现金，他甚至用支票来购买衣服或者其他小商品。

布兰顿让杰米看了他的支票本，告诉她，他可以教她如何平衡收支。杰米之前使用网上银行账户计算存了多少钱和花了多少钱。布兰顿告诉她通过支票本再次核对网上银行账户是个好办法，因为银行或者商店可能会出错。

杰米对支票本上的很多名词的含义都不太清楚。布兰顿给她一一做了解释，“取款”是指从支票账户中取钱，“存款”是指将钱存入支票账户中。当杰米收到银行的回单后，她对账户中的各种存取款，以及她的消费记录便都会一目了然。

利用过去几天保存的收据，杰米开始填写自己的支票本了。她并没有用支票来支付，她现在还只是练习一下。

她的收据包括：

6/03，衣服，96.96美元
6/15，银行存款，20美元
6/16，蛋卷冰激凌，3.50美元
6/21，耳机，15.60美元

她开始填写自己的支票本了，请把那些还没有填写的内容完成：

日期	说明	支出/美元	存款/美元	收支平衡/美元
6/03	衣服	96.96	—	253.04
6/15	家务钱	—	20	273.04

1. 最后一次交易之后，杰米还剩下多少钱呢？

2. 如果杰米写一张150美元的支票，她的账户中是否有足够的钱来支付呢？如果可以支付的话，她还剩下多少钱？

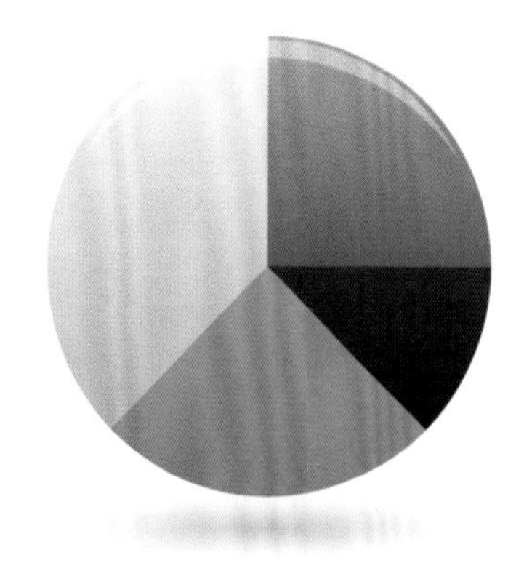

8

使用ATM机

有几次，杰米需要现金，但是只带了借记卡。她和朋友一起去电影院，轮到她买票的时候，收银员告诉她不可以刷卡，因为刷卡的收银系统崩溃了，他告诉杰米目前只能使用现金。

杰米身上没有现金，她知道朋友们随身带的钱也不够。这一刻，她才知道尽管她有很多钱躺在银行里，却没法买票去看场电影。

杰米的好朋友格雷戈告诉她，可以通过ATM机(自动取款机)从账户中取出现金。她只需要把借记卡放入ATM机中就可以取出她想要的数目。格雷戈经常这么做，他带杰米去了剧院边上的ATM机，告诉她如何取钱。还好杰米记得卡的密码，所以她可以在ATM机的键盘上输入密码来完成取现。

使用ATM机取现之前，杰米需要确认账户中是否还有足够的现金。如果杰米取了超额的现金，银行将会收取她一笔30美元的透支费。杰米已经好几周没有去银行了，还有一些现金没来得及去存，可惜放在家里没带在身边。她记得支票账户里还有40美元，所以她取出20美元，然后去售票处买了一张票。售票员给了票和找零。现在她终于可以去看电影了。

杰米的支票账户里实际上还剩26.75美元。她取出了20美元。

1. 她最终会被收取费用吗？

他们买了票后，杰米和朋友们想到还需要买些零食带去剧院。电影票花12美元，还需要12美元买爆米花、饮料和糖果，她准备自己去买回来与格雷戈分享，因为格雷戈的钱用完了。

2. 杰米还需要去自动取款机取钱吗？如果需要的话，需要取多少？自动取款机只提供20美元倍数的纸币。填写杰米可能取出的其他数目，最多到100美元：

3. 20美元
   40美元
   ______美元
   ______美元
   ______美元

杰米以为账户里有40美元，已经取了20美元，应该还剩20美元。实际上有多少？

4. 26.75美元 - 20美元 =

5. 杰米会被收取透支费吗？

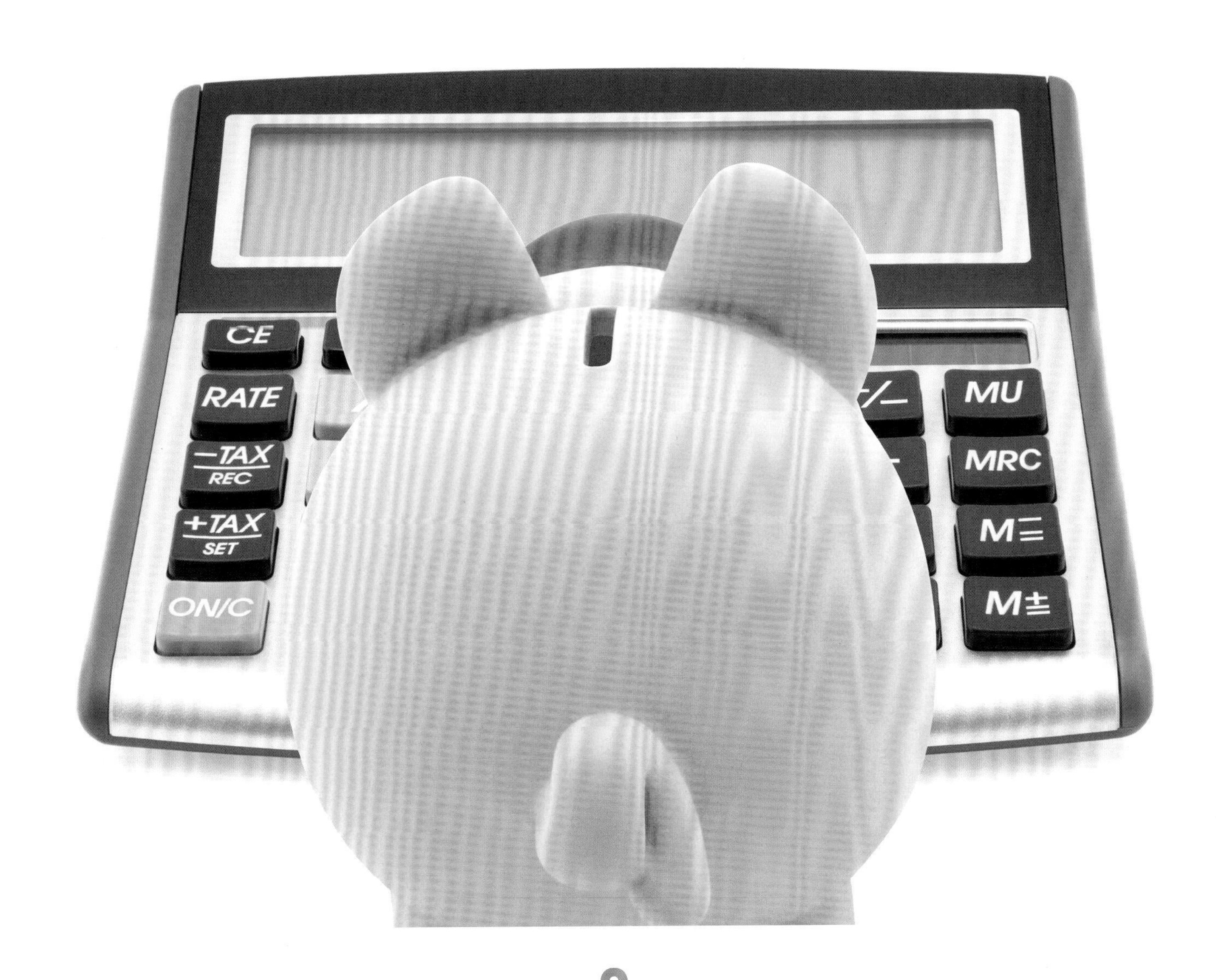

9

预　算

杰米意识到她并没有管理好她的钱。她花了太多的钱，大多数时候都是花了不该花的钱。

她向妈妈求助，因为她知道妈妈非常善于理财。妈妈也很乐意帮助她，她建议杰米做一个预算，并向杰米解释预算其实就是用钱的计划。杰米的预算需要考虑她有多少钱，需要花多少钱，需要留下多少钱。

杰米的妈妈说，预算就像某些规划一样。杰米的新预算会告诉她，什么时候可以花钱，什么时候应该存钱。接下来我们看看杰米和妈妈是如何做预算的。

首先，杰米写下了她定期得到的钱：

10美元，每周的零花钱
25美元，每周做5个小时的零工

然后她写下必须要花的钱：

1美元，每周捐给“食品橱柜”

她每星期存多少钱：

2.50美元，每周零花钱
10美元，家务

杰米的妈妈做了下面的图表，通过它来记录杰米的每周预算。“收入”是杰米得到的所有的钱，“支出”是杰米花掉和存起来的钱。填写下面图表的空白部分：

说明	收入/美元	支出/美元	剩余/美元
零花钱	10	—	10
零花钱剩余	—	2.50	7.50

1. 杰米每星期可以花多少钱？

下周，杰米想给她的朋友玛雅买一份生日礼物。她想给玛雅买一个DVD，会花掉她20美元。

2. 杰米在她的预算中有足够的钱为玛雅买生日礼物吗？她还剩下多少钱？

3. 买了礼物之后，杰米是否有足够的钱再看一场电影？电影票的票价是12美元。她还需要多少钱？

10
如何在购物中做预算

杰米已经攒到够花2个星期的钱了。现在总共有43美元，可以买任何她想要的东西，因为预算告诉她，每个星期可以花21.50美元。

杰米告诉父母，她想做晚饭，感谢他们帮助她管理财务。她想用自己节省下的钱为父母做一顿晚餐。她打算自制调味汁、一份沙拉、一份蒜蓉面包和意大利面。

杰米如何确保不花太多的钱呢？她可以用计算器或手机把所有想买东西的价钱都加起来。杰米认为这将花费太多时间，所以决定仅仅估计杂货店里商品的价格。为了估计价格，她把每一个价格四舍五入到美元为单位，最后将所有的价格加起来。接下来我们看看杰米是否可以负担各种食物的价钱。

将估计值和实际支出填写在下列表格中的空白处。

食物名	价格	四舍五入/美元	实际花费/美元
一盒意大利面	4.99美元/盒	5	4.99
2个番茄（每个重0.25磅）	1.19美元/磅		
一罐番茄酱	0.84美元/罐		
一个洋葱	0.60美元/个		
一头蒜	1.10美元/个		
0.5磅菠菜	3.99美元/磅		
一个辣椒	0.36美元/个		
2个胡萝卜	0.54美元/个		
一瓶意大利汁	2.35美元/瓶		
一个法棍面包	4.50美元/个		
橄榄油	5.69美元/瓶		

1. 用估计方法，杰米有足够的钱来准备这一顿饭吗？（别忘了计算一下两个西红柿和半磅菠菜是多少钱。）

2. 她实际要付多少钱？

11
购物和消费税

杰米去杂货店买一些准备做晚餐的食材，还剩下不少钱。她准备买一件晚餐时穿的衣服，以及妈妈让她买的一些东西。

她去商店买了一些东西，估计费用在预算内。拿到收据的时候，她发现比她预想的要多。为什么会这样呢？原来杰米忘了考虑消费税。消费税是一些额外的花费，是政府对交易行为的收费。消费税由州政府的税收和地方政府（你所在的城镇或城市政府）的税收组成。每个州和城镇都有不同的消费税，所以人们最终支付了0%至11%的消费税。

在杰米所在的州，必须支付8%的消费税。这意味着每一次购物，必须支付8%额外的费用。接下来看一下如何计算消费税。

杰米开始有43美元，现在她在杂货店花了24.69美元，还剩下一些钱：

43美元－24.69美元＝18.31美元

她购买物品的标价如下：

连衣裙，11.99美元
抽纸，1.40美元
电池，3.99美元

如果没有消费税，只需支付约17.40美元，而她还剩下18.31美元，可以足够支付。

8%消费税的另一种说法是杰米的采购成本增加了8%。通过乘法你可以计算8%的成本是多少。

1.

$$X/17.40\text{美元}=\frac{8}{100}$$

$$X=(17.40\text{美元}\times 8)/100$$

$$X=$$

我们也可以换个角度来理解消费税。将8%转换为十进制，将小数点左移两位，8%变成0.08。现在用十进制的乘法来计算杰米的消费：

2.　0.08 × 17.40美元 =

现在把消费税加入到购买成本里，便得到了杰米要支付的消费总额。

3.　杰米的消费在预算之内，还是超过了预算呢？

12

股票和分红

杰米最终了解了储蓄和支票账户的使用窍门。她甚至饶有兴趣地计算预算以及核对购物收据，这有点儿像是一种游戏。

现在，杰米开始着手下一步，弄清楚还能做些什么。一天，她的老师提到了股票。奥尔蒂斯老师解释说，股票（也称为股份）实际上是一个公司的所有权。当你拥有某个公司的股票时，就拥有这家公司的一部分。一家公司可能有成千上万股股票，如果你只有一股，那么你仅拥有这家公司很小的一部分。拥有的股票越多，拥有公司的比例就越大。奥尔蒂斯老师说，当公司赚钱的时候，股票也会变得更值钱。如果这家公司亏损了，那么股票就贬值了。你可以用这种方式从股票市场赚钱。当杰米了解到这些后，她听得更为认真了。奥尔蒂斯老师继续解释，在股票便宜的时候买入，当价格上涨之后再卖掉它们，这样就可以赚一大笔利润。

于是，奥尔蒂斯老师给了同学们一个关于股票的数学项目，希望藉此找出股票市场的运作规律。下面将带你一起去了解一下杰米的项目。

首先，杰米从互联网上或报纸上挑选一家公司的股票，然后假设以当前价格买下它。她选择了一只叫做X公司的股票，价格为76美元/股。

她决定先买5股。

1. 她一共要付多少钱？

76美元/股 × 5股 =

接下来的一周，杰米持续关注这支股票。她将所有的信息添加到下面的图表：

日期	股票价格/(美元/股)	总价值
周一	76	
周二	79	
周三	61	
周四	65	
周五	78	

2. 一周结束后，杰米是赚钱了还是亏钱了？她赚了或是亏了多少？

她在网上查看了一些股票的预测。一位银行家估计她买的股票一年内价格将达到90美元，三年内达到150美元。

3. 从现在起一年后，杰米的5股股票可以值多少钱？

4. 从现在起三年后呢？

5. 如果杰米真的买了股票，她是否应该在一周后出售，还是一年后或者三年后呢？为什么？

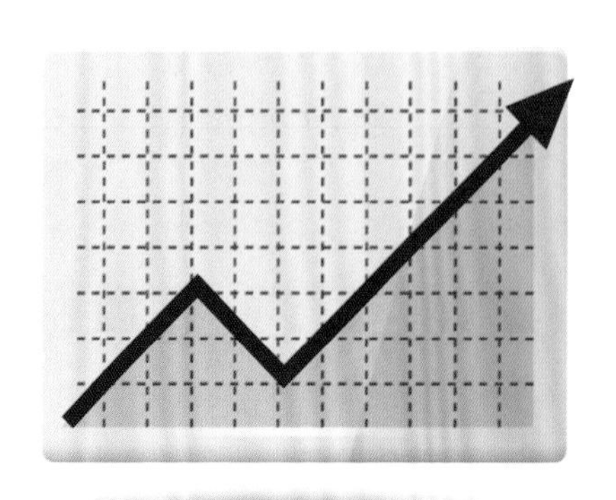

13

如何使用信用卡

杰米之前已经听说过信用卡，但并不知道它究竟能做什么。她对此很感兴趣，于是，她做了一些研究。她查找了网上银行，看看是否能申请一张信用卡。不过她还没有达到申请信用卡的年龄要求。

银行网站上解释了什么是信用卡。信用是对一个人偿还能力的评估。信用卡可以让你购买暂时买不起的东西。如果你现在没有200美元，但需要消费200美元，就可以使用信用卡购买，并在晚些时候偿还这200美元。杰米没有足够的钱，她设想用信用卡给自己买一辆车。

不幸地是，这种想法会给你带来麻烦。刷卡消费并不是免费的，消费的钱最终还是要还的。通常必须在使用信用卡消费的一个月后开始还款。

如果杰米买了一辆车，她将不得不在一个月左右开始还款。她不必一次性偿还所有的费用，但必须还上其中的一部分。与此同时，银行会对剩下还没付清的部分开始收取利息。这可不像储蓄账户的利息可以赚钱——这种利息会被计入杰米要偿还的钱中。还清车款的时间越长，就需要支付更多利息。若要了解其中的原理，我们接着阅读下面的内容。

我们先注意到一些关键的要点：

杰米设想购车费用4000美元。

银行提供的信用卡利率(称为年利率)为每年12%。杰米每个月将被收取12%/12 = 1%的利息。

银行要求信用卡持卡人每月至少还款30美元。

杰米买车后的第一个月内还不需要支付利息。

这里有一个图表，显示了杰米目前的情况。第一行是杰米买车后的第一个月，这个月是不需要支付利息的。她支付 30 美元后，还剩下3700美元没有还清。

第二个月，她同样需要支付30美元。但这个月，她开始需要支付3970美元的利息。3970美元的1%是39.70美元。我们可以看到，她开始遇到麻烦了！她每月需要支付的利息比还款的金额还多近10美元。

试着计算一下第三个月的情况，看看发生了什么。

月份	起始金额/美元	每月还款额/美元	利息/美元	还需要还款额/美元
1	4000	30	0	3970
2	3970	30	39.70	4009.70
3				

1. 和前一个月相比，杰米需要偿还更多吗？多多少呢？

2. 杰米用信用卡买车的想法可行吗？为什么？

3. 如果杰米每月还款500 美元的话，情况会怎么样？她能够最终还清信用卡吗？

14

货币兑换

杰米一家准备去墨西哥旅行！杰米非常兴奋，开始为这次旅行存钱了。

杰米的哥哥约书亚告诉她墨西哥人用的钱叫比索，而不是美元。那杰米怎么才能购买东西呢？约书亚告诉她，可以到银行把美元换成比索，1美元不等于1比索。她需要按照汇率换算，把美元兑换成等值的比索。

美元与比索的汇率是1:12.39。符号“:”表示1美元可以兑换12.39比索。我们也可以反过来换算，1比索可以兑换成多少美元。比索对美元的汇率是0.08:1 或者说每1比索等于0.08 美元。

如果杰米存了50美元，她在银行能兑换多少比索？

1. 50美元 × 12.39 =

杰米可能不会花掉所有的比索。她把50美元都换成了比索，但是现在还剩了90比索，她想要换回美元。

2. 如果换算成美元，应该是多少钱？

然而，一旦杰米去银行兑换，她不会得到预想的数目，因为银行会收取一定的服务费。

3. 如果银行要收取兑换金额10%的服务费，那么实际上杰米可以拿到多少比索。首先弄清楚杰米还剩多少美元，然后将其换算成比索。

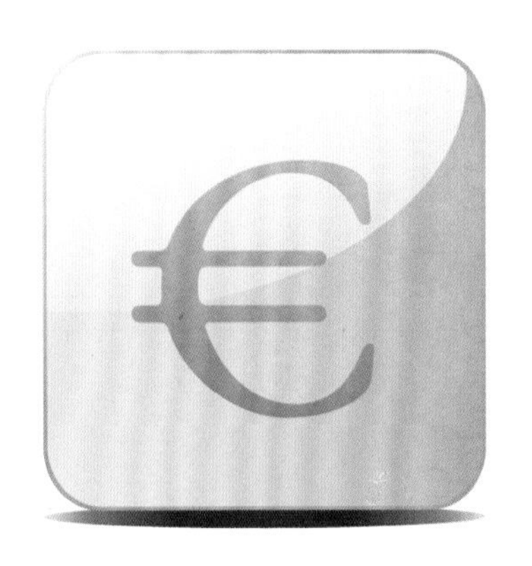

15
小 结

杰米从家人和朋友那里学到了很多金融知识。现在她有了支票账户和储蓄账户，做了预算，并计划买一些股票。

看看你是否记得杰米到目前为止已学到的东西。

1. 如果你的零花钱最近提高至每周15美元，一年能获得多少？

2. 你打算把15美元的三分之一存起来，并把它存入一个储蓄账户。每周可以存多少？

3. 你的储蓄账户每月收入 0.5%的利息。如果银行账户中有 600 美元，那么一年后你可以得到多少利息？

4. 你用借记卡买一个45美元的视频游戏，并往支票账户上存入36 美元。借记卡里原来有68.50 美元，而账户中至少要剩 10 美元以免支付小额管理费。那么现在需要支付小额管理费吗？

5. 你的预算是每周存下 15 美元，支出 12 美元。你想要买的一些美术用品需要11.50美元，但你忘记了考虑消费税。如果购买所有预算中的东西，并算上7%的消费税，还能保证不超出预算吗？

6. 你买了每股40 美元的股票，现值 47 美元。如果你买了 32 股，那么现在一共赚多少钱？

7. 你的信用卡要支付24%的年利率，那么每月的利率是多少？

8. 如果你用信用卡支付了100美元，并不马上还清的话，那么在原来成本的基础上，月底最终需要支付多少额外的利息？

9. 如果在一年内没有还清100 美元的话，那么年底要支付多少利息？

10. 如果去法国旅游，那 美元换成欧元。当到达巴黎时，你会看到一个牌子，上面写着“欧元兑美元的汇率是1:1.30”。如果把50美元兑换成欧元，你将得到多少欧元(假设没有额外的兑换费用)？

参考答案

1.

1. 多了（520美元 - 104美元 = ）416美元
2. 10美元 ÷ 4 = 2.50美元
3. 10美元 - 2.50美元 - 1美元 = 6.50美元

2.

1. 75美元
2. 75美元
3. (4周 × 2.50美元/周) + 75美元 = 85美元

3.

1. 3小时 × 5美元/小时 = 15美元，40美元 ÷ 5美元/小时 = 8小时
2. 他必须每周挣15美元，所以必须每周工作3小时
3. 2小时
4. 100 ÷ 6美元/小时 = 16 2/3 小时
5. 由于需要上学，他可能没有那么多时间

工作时间（邻居2）/小时	报酬/（6美元/小时）
1	6
2	12
3	18

工作时间（邻居3）/小时	报酬/（12美元/小时）
1	12
2	24
3	36

4.

1. 216美元
2. 多了，多了76美元
3. 60美元
4. 200美元
5. 100美元
6. 多了，多50美元
7. 第2种，因为她有能力保证最低存款的要求，同时第1种会让她支付更多的费用

5.

1. 0.42美元
2. 141.27美元
3. 141.40美元
4. 501.50美元

6.

1. 29.99美元 +(2 × 15.99美元)+ 34.99美元 = 96.96美元
2. 27.7
3. 不,没有超过一半

7.

1. 253.94美元
2. 是的,她有103.94剩余

日期	说明	支出/美元	存款/美元	收支平衡/美元
6/03	衣服	96.96	—	253.04
6/15	家务钱	—	20	273.04
6/16	蛋卷冰激凌	3.50	—	269.54
6/21	耳机	15.60	—	253.94

8.

1. 不会
2. 是的,她还需4美元
3. 60, 80, 100
4. 6.75美元
5. 会收取 (6.75美元 - 20美元 < 0)

9.

1. 21.50美元
2. 是的, 她还剩下1.50美元
3. 没有钱看电影了， 她还差10.50美元

说明	收入/美元	支出/美元	剩余/美元
零花钱	10	—	10
零花钱剩余	—	2.50	7.50
家务钱	25	—	32.50
家务钱剩余	—	10	22.50
捐款	—	1	21.50

10.

1. 是的, 够用
2. 24.96美元

食物名	价格	四舍五入/美元	实际花费/美元
一盒意大利面	4.99美元/盒	5	4.99
2个番茄（每个重0.25磅）	1.19美元/磅	1	0.60
一罐番茄酱	0.89美元/罐	1	0.89
一个洋葱	0.60美元/个	1	0.60
一头蒜	1.10美元/个	1	1.10
0.5磅菠菜	3.99美元/磅	2	2.00
一个辣椒	0.76美元/个	1	0.76
2个胡萝卜	0.54美元/个	1	1.08

续表

一瓶意大利汁	2.75美元/瓶	3	2.75
一个法棍面包	4.50美元/个	5	4.50
橄榄油	5.69美元/瓶	6	5.69

11.

1. 1.39美元
2. 1.39美元
3. 她花的比预算多(17.40美元+1.39美元=18.79美元)

12.

日期	股票价格/(美元/股)	总价值
周一	76	380
周二	79	395
周三	61	305
周四	65	325
周五	78	390

1. 380美元
2. 她赚了10美元
3. 90美元/股 × 5股=450美元
4. 150美元/股 × 5=750美元
5. 三年之后卖出，因为三年后她的股票价值最高

13.

1. 是的,她要多还40.10美元
2. 不可行,由于需要支付利息，她要偿还比4000美元要多的钱
3. 是的,她能还清信用卡欠款

月份	起始金额/美元	每月还款额/美元	利息/美元	还需要还款额/美元
1	4000	30	0	3970
2	3970	30	39.70	4009.70
3	4009.70	30	40.10	4049.80

14.

1. 619.50比索
2. 90比索 × 0.08 = 7.20美元
3. （1-10%）× 50美元 = 45.00美元，45 × 12.39 = 557.55比索

15.

1. 52 × 15美元 = 780美元
2. 15/3美元 = 5美元
3. 总金额计算公式 = $600(1+0.005/12)^{12\times1}$美元
   总金额 = 603.01美元
   利息计算公式= 总金额 - 本金
   利息 = 603.01美元 - 600美元
   利息 = 3.01美元
4. 不需要，你账号的钱比10美元多（68.50美元 - 45美元 + 36美元 = 59.50美元）
5. 会的，你会超出预算0.31美元（0.07 × 11.50美元 = 0.81美元，11.50美元 + 0.81美元 = 12.31美元）
6. 多赚（7美元/股 × 32股= ）224美元
7. 2%
8. 2美元
9. 24美元
10. 50美元 ÷ 1.30 = 38.46欧元

INTRODUCTION

How would you define math? It's not as easy as you might think. We know math has to do with numbers. We often think of it as a part, if not the basis, for the sciences, especially natural science, engineering, and medicine. When we think of math, most of us imagine equations and blackboards, formulas and textbooks.

But math is actually far bigger than that. Think about examples like Polykleitos, the fifth-century Greek sculptor, who used math to sculpt the "perfect" male nude. Or remember Leonardo da Vinci? He used geometry—what he called "golden rectangles," rectangles whose dimensions were visually pleasing—to create his famous *Mona Lisa*.

Math and art? Yes, exactly! Mathematics is essential to disciplines as diverse as medicine and the fine arts. Counting, calculation, measurement, and the study of shapes and the motions of physical objects: all these are woven into music and games, science and architecture. In fact, math developed out of everyday necessity, as a way to talk about the world around us. Math gives us a way to perceive the real world—and then allows us to manipulate the world in practical ways.

For example, as soon as two people come together to build something, they need a language to talk about the materials they'll be working with and the object that they would like to build. Imagine trying to build something—anything—without a ruler, without any way of telling someone else a measurement, or even without being able to communicate what the thing will look like when it's done!

The truth is: We use math every day, even when we don't realize that we are. We use it when we go shopping, when we play sports, when we look at the clock, when we travel, when we run a business, and even when we cook. Whether we realize it or not, we use it in countless other ordinary activities as well. Math is pretty much a 24/7 activity!

And yet lots of us think we hate math. We imagine math as the practice of dusty, old college professors writing out calculations endlessly. We have this idea in our heads that math has nothing to do with real life, and we tell ourselves that it's something we don't need to worry about outside of math class, out there in the real world.

But here's the reality: Math helps us do better in many areas of life. Adults who don't understand basic math applications run into lots of problems. The Federal Reserve, for example, found that people who went bankrupt had an average of one and a half times more debt than their income—in other words, if they were making $24,000 per year, they had an average debt of $36,000. There's a basic subtraction problem there that should have told them they were in trouble long before they had to file for bankruptcy!

As an adult, your career—whatever it is—will depend in part on your ability to calculate mathematically. Without math skills, you won't be able to become a scientist or a nurse, an engineer or a computer specialist. You won't be able to get a business degree—or work as a waitress, a construction worker, or at a checkout counter.

Every kind of sport requires math too. From scoring to strategy, you need to understand math—so whether you want to watch a football game on television or become a first-class athlete yourself, math skills will improve your experience.

And then there's the world of computers. All businesses today—from farmers to factories, from restaurants to hair salons—have at least one computer. Gigabytes, data, spreadsheets, and programming all require math comprehension. Sure, there are a lot of automated math functions you can use on your computer, but you need to be able to understand how to use them, and you need to be able to understand the results.

This kind of math is a skill we realize we need only when we are in a situation where we are required to do a quick calculation. Then we sometimes end up scratching our heads, not quite sure how to apply the math we learned in school to the real-life scenario. The books in this series will give you practice applying math to real-life situations, so that you can be ahead of the game. They'll get you started—but to learn more, you'll have to pay attention in math class and do your homework. There's no way around that.

But for the rest of your life—pretty much 24/7—you'll be glad you did!

1 ALLOWANCE MATH

Jamie used to get an allowance every week of $2. She usually spends half of it and saves the other half. In fact, she has two piggy banks—one for spending and one for saving. However, Jamie just got a raise in her allowance. Her parents told her she has used her money wisely, and because she's getting older, they will give her more money each week. Her parents have decided to give her $10 a week.

Jamie likes going shopping, so she's excited she's getting more money. She also wants to save some of her allowance so she can buy more expensive things later on. Finally, Jamie likes helping other people. She wants to give some of her allowance away to charity.

Jamie knows $10 is a lot more than $2, but she's not sure just how it all adds up. How much more will she be making in allowance per year?

First you need to find out how much she was making when she got $2 in allowance. There are 52 weeks in a year, so:

52 weeks x $2 a week = $104

Now she makes $10 a week:

52 weeks x $10 a week= $520

1. Jamie can see she's making a lot more money, but just how much more is she making? You do the math.

Jamie decides to save one-fourth of her money. One-fourth is like dividing something into four parts. She will be dividing her allowance into four parts, and putting one part in her savings piggy bank.

Just divide her weekly allowance by 4.

2. How much will she save each week?

She has three-fourths of her allowance left. She wants to give $1 a week to a food pantry to help people.

3. Now how much money does she have left to spend every week, after figuring in her savings and donations?

2 GETTING GIFTS

Jamie's birthday is in a week. She keeps getting birthday cards in the mail from all her friends and family. One card arrives from her grandma with a check inside. Her grandma has given her $50. Her parents also give her an early birthday present. They see how good Jamie is at saving money, and they want to add to her savings. They give her a check for $40, but they tell her she has to add it to her savings piggy bank. Finally, Jamie's uncle sends her yet another check for $35, and he tells her she should use it to buy something she really wants for her birthday.

Suddenly, Jamie has a lot more money than she did before. She takes all of her checks to the bank and cashes them in for $125 total. Now she has to decide how she wants to use it all.

Jamie got three checks for $50, $40, and $35. Two of the checks have rules that go along with them: the $40 check she can only use for savings, not spending. And the $35 check she should use for buying something she wants. Jamie really wants to buy a new video game that costs $50. The $35 doesn't quite cover it. How much more does she need?

$$\$50 - \$35 = ?$$

She does the math and sees she needs just $15 more. She can't use the check from her parents, which is for saving. She can use the check her grandma sent, though.

1. How much will she have left after she buys her video game? You can either do the math step by step:

   spending uncle's check + spending some of grandma's check
   + saving parent's check = ?

   $$(\$35 - \$35) + (\$50 - \$15) + \$40 =$$

   Remember, do the math in the parentheses first.

2. Or you can look at the big picture:

$125 total – $50 for the video game =

It turns out, Jamie has $75 total to add to her savings. She already has four weeks' worth of allowance savings, too.

3. How much savings does she have in all? Turn back to section 1 to see how much Jamie saves each week.

3 EARNING MONEY

Jamie is really excited about all the money she's getting these days. She tells her friend Kareem. He tells her he also wants to make some money so he can pay for a bike he's been wanting. Jamie and Kareem talk it over and decide they want to start doing some work to earn more money.

Jamie wants to do errands around the house. Her dad just started working full time, so he has less time to clean up and do laundry. Jamie asks her dad if there are any chores she can do in return for pay. Her dad thinks for a bit, and tells her she could pull weeds outside, do a load of laundry every week, and help him and her mom clean the house once in a while. They'll pay Jamie $5 an hour for doing these errands.

Meanwhile, Kareem decides he wants to walk dogs for a little money. He knows all his neighbors, and lots of his neighbors have dogs. He once agreed to walk one of his neighbor's dogs a few months ago, and she paid him. Kareem realizes he could take the **initiative** and ask his neighbors if they need him to walk their dogs. He asks around and finds three neighbors who agree he can walk their dogs once a week. One will pay him $5 every time, one will pay him $6, and the last will pay him $12 because he has two very big dogs.

Jamie and Kareem are both making money now. How does it add up?

Jamie is doing three hours of chores each week.

1. How much money is she making? How many hours a week would she have to work if she wanted to make $40?

Kareem doesn't just earn one flat rate. He earns different amounts of money depending on which neighbor he is working for. Fill out the charts below to figure out how much he makes for each hour he works. The first chart is filled out for you.

Hours worked for Customer #1	Money earned at $5/hour
1	$5
2	$10
3	$15

Hours worked for Customer #2	Money earned at $6/hour
1	
2	
3	

Hours worked for Customer #3	Money earned at $12/hour
1	
2	
3	

2. How many hours does Kareem have to work for Customer #1 to make as much as Jamie?

3. How many hours would he have to work for Customer #3 to make more than Jamie?

4. How many hours would he have to work for Customer #2 to earn $100 a week?

5. Would he realistically have time to work that many hours during the school year?

4 OPENING A SAVINGS ACCOUNT

Now Jamie has a lot of money because she's been saving so well. Her piggy bank doesn't have enough room anymore for her savings. It's time for her to open a savings account at the bank.

Her dad explains to her that savings accounts are safe places to keep your money. A savings account will keep her money safe but out of sight, so she's not tempted to use it when she doesn't really need to.

Jamie and her dad go to the bank and talk to a bank employee about what sort of savings account Jamie should open. The bank offers a few choices. Some of them have monthly fees, while others require you to have at least a certain amount of money in the account at all times. Jamie may also have to pay a fee if she doesn't use her account often enough.

Jamie has to compare the different types of savings accounts at the bank to see which one will be best. First, she looks at a savings account that will make her pay a monthly fee. The fee is $10 per month, but will change to $8 a month after the first year.

Calculate how much she will pay the bank over two years:

1. ($10 x 12 months) + ($8 x 12 months)=

2. Is this amount more or less than Jamie has in savings? If it's more, how much more?

The other savings account she could get has to have at least $50 in it at all times. If the account has less than $50, she'll have to pay $30. Right now, she has $140 from her allowance savings, birthday checks, and chores. She's planning on continuing to do chores 3 hours a week at $5 an hour. She also is planning on buying a TV for her room next month (in four weeks), which will cost $100. Will she have enough savings left four weeks from now so she won't have to pay the fee?

Figure out how much she will make in chores over four weeks:

3. 3 hours x $5 an hour x 4 weeks =

4. Now add that to her savings she has already. How much does she have?

5. How much will she have after she buys the TV?

6. Is it more or less than the $50 minimum for the savings account? By how much more?

7. Based on your answers, do you think Jamie should get the first savings account or the second? Why?

5 INTEREST RATES

At the bank, Jamie opens up her new savings account. The bank employee also tells her that her savings account will earn her something called interest. Interest is extra money she will earn just for having money in the bank. The more money is in a savings account, the more money she'll earn in interest. And then she gets to earn interest on her interest, which equals even more money. Jamie wishes she had put her money in the bank before now. At home in the piggy bank, her money didn't earn any interest.

When she gets home, she asks her dad more about interest. She wants to know how much she'll make in interest, based on how much she has in the bank. Her dad tells her she won't make a lot, but she'll make a little bit. You'll see how to do it on the next page.

The math equation you use to calculate how much money you will have after interest is:

Total amount = $P\,(1 + r/n)^{nt}$

That may look complicated, but it's not too bad once you know what all those letters mean.

- P is how much money you have in a savings account
- r is the rate of interest, which the bank tells you
- n is the number of times interest is calculated over time. It might be calculated once a month or four times a year, or once a year.
- t is the number of years you're looking at

Here's what Jamie's interest equation looks like:

P= $140 in her savings account
r= .3 percent, or .003. That is another way of saying one-third out of a hundred
n= 12, because her interest is calculated once a month for twelve months a year
t= 1 year

Now you just plug in the numbers to find out how much interest she would get if she left her $140 in the account for a year. The little numbers at the end that are raised up are called exponents, and can also be written out ^nt. You can use a calculator to punch in the exponents.

Total amount = $\$140\,(1 + .003/12)^{12 \times 1}$
Total amount = $140.42

How much interest did Jamie earn?

1. $140.42 – $140.00 =

2. How much money would Jamie have in the bank if she left her $140 for three years?

3. How much money would she have after a year if her interest rate were 1% (.01)?

4. How much money would she have after interest if she started out with $500?

6 CHECKING ACCOUNT MATH

Jamie has saved up a lot of money in her savings account after a few months. She also keeps some spending money at home, so she has it nearby when she wants to buy something.

It's hard to keep track of all that spending money, though. Jamie once lost a $20 bill that she got for doing chores. She looked all around the house for it, but she never found it. She also finds money all over her room—under her bed, in her backpack, and even in the corners of her closet.

Jamie is talking about her problem with her friend Maya at school. Maya tells her she used to have the same problem, but then she opened a checking account at her bank. She also got a debit card to go along with her checking account. A debit card is sort of like cash, Maya explains to Jamie. You use it at stores to spend money in your checking account.

Maya's checking account is separate from her savings account, which she also has at the same bank. She doesn't normally take money out of her savings account. She's saving that money for college later on. However, she does use money out of her checking account all the time, and she puts more back in when she gets spending money. Instead of using dollar bills and coins, she just uses a card to pay for things she buys.

Jamie heads back to the bank with her dad and opens a checking account. Will Jamie be able to keep track of how much money is in her checking account? Will she spend more money because she's using a card to pay for things instead of cash?

After saving up for several months, Jamie now has plenty of spending money. She has $350 all together, even after buying a few things like that new video game. Now that she has a shiny new debit card, she wants to try it out. She goes to the store and buys some clothes, and doesn't really pay attention to the price tags. This is what she buys:

Jeans, $29.99
2 T-shirts, $15.99 each
Shoes, $34.99

1. How much did she spend? Don't forget to multiply the T-shirt price by 2.

How much money does Jamie have left in her checking account? She started out with $350 and she spent $96.96, so she has $253.04 left.

Jamie spent quite a bit of money at the store. You can figure out what percent of her spending money she spent. Percents are parts out of a hundred. For example, if Jamie had $100 and she spent $40, she would have spent 40% of her money. However, Jamie has more than $100, so we want to find out what percent she spent.

You'll have to do some cross multiplication:

$96.96/$350 = X/100

Solve for X:

2. $96.96 x 100 = $350 x X
   X= ($96.96 x 100)/$350
   X= _____%

3. Is the percent of the money Jamie spent more than half of her checking account? (Half would be 50%.)

7 BALANCING A CHECKBOOK

Jamie also received a brand-new checkbook when she opened her checking account. She's not entirely sure what to do with it, but she wants to learn. Her older cousin Brandon is visiting, so she asks him about checks.

Brandon tells Jamie he uses checks all the time. He has an apartment and pays his rent by sending checks to his landlord. He also uses checks to pay for his electricity and heat bills he pays every month. Once in a while, he even uses a check to pay for clothes or other small purchases because he forgot his debit card and he doesn't have any cash with him.

Brandon pulls his checkbook out to show Jamie. He tells her he's going to teach her how to balance a checkbook. She has been using her online banking account to figure out how much money she has and how much she's spent. Brandon tells her it's a good idea to double check the online account, in case the bank or a store made a mistake. That's where balancing a checkbook comes in.

Jamie has a lot of questions about what the different words in the checkbook mean. Brandon explains a "withdrawal" is taking money out of a checking account. A "deposit" is adding money to a checking account. The "balance" is how much is in the checking account in total. Jamie will know what she's done with her account because she can look at all the receipts she gets from the bank and whenever she buys things.

Jamie starts to fill in her own checkbook using the receipts she has saved over the past few days. She hasn't used checks to pay for them yet, but she's just practicing for now.

She has receipts for:

6/03, Clothes, $96.96
6/15, Bank deposit, $20
6/16, Ice cream cone, $3.50
6/21, Headphones, $15.60

She starts to fill in the checkbook. Finish what hasn't been filled out yet.

Date	**Description**	**Withdrawal**	**Deposit**	**Balance**
6/03	Clothes	$96.96	—	$253.04
6/15	Chore money	—	$20	$273.04

1. How much money does Jamie end up with after her last transaction?

2. If Jamie wrote a check for $150, would she have enough money in her checking account? If so, how much would she have left?

8 USING AN ATM

Sometimes Jamie wants cash, but all she has is her debit card. One day, she's out with her friends at the movie theater. As she moves up in line, the cashier tells her she can't use her debit card because the system is broken that allows the cash registers to accept cards. He tells her she can still pay with cash, though.

Jamie doesn't have any cash, and she knows her friends don't carry much money around so she can't borrow from them. For a minute, she thinks she can't go to the movies, even though she has plenty of money sitting in the bank.

Then Jamie's friend Greg tells her she can use an ATM (Automated Teller Machine) to get cash out from her checking account. All she has to do is swipe her debit card in an ATM and withdraw however much she needs. Greg has done it plenty of times, so he leads Jamie over to the theater's ATM to show her how to get money. Fortunately, Jamie remembers her PIN, because she needs to tap that on the keyboard in order to use the ATM.

Jamie also has to be sure she has enough money in her checking account before she can take money out using the ATM. Jamie's bank will charge her an overdraft fee of $30 if she takes out too much money. Jamie hasn't been to the bank in a few weeks. She has some money sitting at home (which she forgot to bring to the theater) that she needs to deposit, but she hasn't gotten around to it yet. She thinks she has $40 in her checking account, so she takes out $20, goes back to the cashier, and hands him the money. He gives her change and a ticket, and she's all ready to see the movie.

Jamie really had $26.75 left in her checking account. She took out $20.

1. Did she end up getting charged the fee?

After they buy their tickets, Jamie and her friends realize they want snacks to take in to the theater. Their tickets were $12. Jamie needs $12 more to buy popcorn, drinks, and candy—she's buying some herself and sharing it with Greg, who ran out of money.

2. Will Jamie need to take out more money from the ATM? If so, how much does she need?

The ATM only gives out money in multiples of $20. Fill in the rest of the amounts Jamie could take out, up to $100:

3. $20
   $40
   $_____
   $_____
   $_____

Jamie thinks she had $40 in her account, and she withdrew $20. She thinks she has $20 more. How much does she really have?

4. $26.75 – $20 =

5. Will Jamie get charged an overdraft fee?

9 BUDGETING

Jamie realizes she hasn't been managing her money very well. She's spending too much money, and she isn't spending it wisely all the time.

She asks her mom for help, because she knows her mom is good at saving money. Her mom is happy to help, and she suggests that Jamie make a budget. A budget, she explains to Jamie, is a written plan for using money. Jamie's budget will take into account how much money she has, how much money she needs to spend, and how much money she has left over.

A budget is kind of like a map, Jamie's mom says. Jamie's new budget will tell her when she can spend money, and when she should save. Check out the next page to see how Jamie and her mom make a budget.

First, Jamie writes down all the ways she regularly gets money:

$10 weekly allowance

$25 doing 5 hours of chores a week

Then she writes down when she absolutely needs to spend her own money:

$1 a week to the food pantry

And how much she saves each week:

$2.50 from her weekly allowance
$10 from chores

Jamie's mom makes the following chart, which will keep track of Jamie's weekly budget. "Income" is all the money Jamie makes. "Expenses" are the ways Jamie spends and saves money. Fill in the rest of the chart.

Description	Income	Expense	Available
Allowance	$10	—	$10
Allowance savings	—	$2.50	$7.50

1. How much spending money does Jamie have every week?

Next week, Jamie wants to buy her friend Maya a birthday present. She was thinking about getting her a DVD that costs $20.

2. Does Jamie have enough room in her budget for Maya's birthday present? How much does she have left over?

3. After buying the present, will Jamie have enough to see another movie at the theater, which costs $12? How much more would she need?

10 SHOPPING MATH: PUTTING A BUDGET INTO PRACTICE

Jamie has been saving up her spending money for two weeks. Now she has $43 total to spend on whatever she wants, since her budget tells her she can spend $21.50 a week.

Jamie tells her parents she wants to make them dinner, to thank them for helping her with her **finances**. She wants to pay for the whole dinner herself, and has saved up her spending money to do so. She plans on making spaghetti with homemade sauce, a salad, and garlic bread.

How will Jamie make sure she doesn't spend too much money? She could add up everything she wants to buy on a calculator or her phone. Jamie thinks that will take too much time, so she decides to just estimate prices at the grocery store. To estimate, she rounds each price up to the nearest dollar. Then she adds all the prices together. Turn to the next page to see if Jamie can afford everything for her meal.

Finish filling out the chart with estimates and how much each item will actually cost Jaime at the register.

Item	Price	Estimate	Actual cost
1 box spaghetti	$4.99 a box	$5	$4.99
2 tomatoes(each tomato weighs one quarter of a pound)	$1.19 a pound		
1 can tomato paste	$.89		
1 onion	$.60 each		
1 head garlic	$1.10 each		
½ pound spinach	$3.99 pound		
1 pepper	$.76 each		
2 carrots	$.54 each		
1 bottle Italian dressing	$2.75 per bottle		
1 loaf French bread	$4.50		
olive oil	$5.69 per bottle		

1. Using estimations, does Jamie have enough money to buy everything for her meal? (Don't forget to calculate how much money two tomatoes and half a pound of spinach are.)

2. How much will she actually pay at the register?

11 SHOPPING AND SALES TAX

Jamie has plenty of money left over from her grocery store trip to buy something else for dinner. She thinks it would be nice if she bought a new dress for dinner and a couple things her mom asked her to pick up.

She goes to the store and gets a few things, estimating that she's staying under budget. When she gets the receipt, though, she sees it's more than she added up in her head. What's going on?

Jamie forgot to take sales tax into account. Sales tax is a little extra money the government charges us for making purchases. A sales tax is made up of two parts: the tax created by the state government, and the tax created by the local government (the government that runs your town or city). Every state and town has a different sales tax, so people end up paying between 0% and 11% in sales tax.

In Jamie's state, she has to pay 8% sales tax. That means for every purchase she makes, she has to pay 8% more. See how to calculate sales tax on the next page.

Jamie started out with \$43, and now she spent \$24.69 at the grocery store. She has some money left:

$\$43 - \$24.69 = \$18.31$

Her purchases are listed below, according to the prices on the tags:

Dress, \$11.99
Tissues, \$1.40
Batteries, \$3.99

Without sales tax, Jamie thought she was going to pay about $17.40, which was less than the $18.31 she had left to spend.

The 8% sales tax is another way of saying 8 one-hundredths of the cost of Jamie's purchases are added to the receipt. Cross-multiplication will tell you just how much 8% of the cost is.

1. X/$17.40=8/100
   X= ($17.40 x 8)/100
   X=
   You can also figure out the sales tax a different way. To **convert** 8% to a decimal, move the decimal point over two places to the left. So 8% becomes 0.08. Now multiply the decimal by Jamie's purchases:

2. 0.08 x $17.40 =

Now add the sales tax to the purchase cost to get the amount of total money Jamie will have to pay.

3. Is Jamie still following her budget, or has she spent more money than her budget allowed?

12 STOCKS AND BONDS

Jamie has finally gotten the hang of her savings and checking accounts. She even has fun figuring out budgets and adding up receipts—it's kind of like a game.

Now Jamie wants to take the next step and figure out what else she can do financially. One day in school, her teacher mentions stocks. Ms. Ortiz explains that stocks (also called shares) are actually pieces of ownership in a company. When you own a stock, you partially own a company. A company may have thousands of shares, so if you only own one stock in the company, you only own a tiny piece of that company. The more stocks you own in a company, the bigger your piece of ownership. As a company makes money, your stocks are worth more money, Ms. Ortiz says. If a company loses money, your stocks are worth less money. You can use these ideas to make money from the stock market.

Jamie listens even more closely when she hears that. Ms. Ortiz goes on to explain. By buying shares when they are cheap, and waiting to sell them until they are expensive, you can make a big profit.

Then Ms. Ortiz gives the class a math project to do on stocks, to figure out just how the stock market works. The next page will guide you through Jamie's project.

First, Jamie has to pick from the Internet or a newspaper a company with stock, and then imagine she's buying it for whatever price it's selling at. She picks a stock from Company X for $76. She also decides to buy 5 shares.

1. How much will she pay in total?

$$\$76 \times 5 =$$

Next Jamie follows her stock for a week. She adds all her information to the chart below:

Day	**Stock Price**	**Total Value of Shares**
Monday	$76	
Tuesday	$79	
Wednesday	$61	
Thursday	$65	
Friday	$78	

2. Did Jamie make or lose money by the end of the week? How much did she make or lose?

She checks some stock predictions online. One banker estimates her stock will be worth $90 in a year, and $150 in three years.

3. How much would Jamie's 5 stocks be worth a year from now?

4. How about three years from now?

5. If Jamie were to really buy stocks, should she sell them at the end of the week, a year from now, or three years from now? Why?

13 CREDIT CARD MATH

Jamie has heard about credit cards before, but she doesn't know exactly what they are. She's interested, though, so she does some research. She looks up her bank online to see if she could get a credit card. It turns out she's not old enough yet to apply for one.

The bank website explains what a credit card is. Credit is trust in someone's ability to pay back money. Credit cards let you buy things you can't afford right away. If you don't have $200 right now, but you need to spend $200, you can use a credit card to make the purchase and pay the $200 later. Jamie imagines herself buying a car with her credit card, even though she doesn't have nearly enough money.

Unfortunately, that kind of thinking can get you in trouble. Credit isn't just free money. You still have to eventually pay for what you buy. You usually have to at least start paying for your credit card purchases a month later.

If Jamie were to buy a car, she would have to start making payments after a month or so. She wouldn't have to pay for the whole thing at once, but she would have to pay a little bit. Meanwhile, the rest of the money she hasn't paid off starts earning interest. This isn't like interest on a savings account, where you earn money—this interest adds money Jamie will have to pay back. The longer she waits to pay off the entire cost of the car, the more interest she'll have to pay. To see how it works, check out the following page.

Here are some key facts:

Jamie's imaginary car costs $4,000.

The bank offers a credit card with an interest rate (called an Annual Percentage Rate) of 12% a year. Each month, Jamie will be charged 12%/12 months = 1%.

The bank requires credit card holders to pay at least $30 each month toward their purchases.

Jamie doesn't start having to pay interest for a month after she makes her purchase.

Here's a chart that shows Jamie's situation. The first row is the very first month after Jamie buys her car. She has a month before she has to start paying interest. She ends up with $3,700 left to pay off after she pays $30.

The next month, she pays another $30. But this month, she starts having to pay interest on that $3,970. 1% of $3,970 is $39.70. You can see she starts to run in to trouble here! The interest she has to pay is almost $10 more than the payments she makes each month.

Do one more month to see what happens.

Month	Original Amount	Amount Paid	Interest	Still to Be Paid
1	$4,000	$30	0	$3,970
2	$3,970	$30	$39.70	$4,009.70
3				

1. Does Jamie have even more to pay off than she did the month before? How much more?
2. Does it make sense for Jaime to buy a car using a credit card? Why or why not?
3. What if Jamie paid $500 a month? Would she be able to pay off her credit card eventually?

14 EXCHANGING MONEY

Jamie's family just told her they are going on a vacation to Mexico! Jamie is very excited, and she starts saving up her money for the trip.

Jamie's older brother Joshua tells her people use pesos as money in Mexico. They don't use dollars. How will Jamie be able to buy anything? Joshua tells her she can take her money to the bank and exchange it for pesos. However, he says, one U.S. dollar doesn't equal one Mexican peso. She'll have to use an exchange rate to figure out how many pesos she'll be able to get with her money.

The exchange rate between dollars and pesos is 1:12.39. The ":" tells you that for every dollar, you'll get 12.39 pesos. You can also flip it around and think about how many dollars one peso is worth. The peso-to-dollar exchange rate is 0.08:1, or $0.08 for every peso.

If Jamie has $50 saved up, how many pesos can she get at the bank?

1. $50 x 12.39 =

Maybe Jamie won't spend all her money in pesos. She exchanged all $50 for pesos, but now she has 90 pesos left, and she wants to change it back into U.S. dollars.

2. How much is that in U.S. dollars?

However, once Jamie goes to the bank to exchange her money, she doesn't actually get the number of pesos she thought she was going to get. The bank takes some money as a fee for doing the exchange.

3. How much money in pesos will Jamie have if the bank takes 10% of the U.S. money she brings to the bank to exchange? First figure out how many dollars Jamie will have left, and then convert to pesos.

15 PUTTING IT ALL TOGETHER

Jamie has learned a lot from her family and friends about banking. Now she has a checking account and savings account, a budget, and is thinking about buying some stocks.

See if you can remember some of the things Jamie has learned so far.

1. If your allowance was recently raised to $15 a week, how much will you make in a year?

2. You want to save one-third of your $15 allowance and put it into a savings account. How much will you save each week?

3. The savings account you opened earns 0.5% interest every month. How much interest will you make in a year if you have $600 in the bank?

4. You buy a new video game with your debit card for $45. Then you deposit $36 in your checking account. You started out with $68.50. You need a balance of at least $10 in your account to avoid paying a fee. Will you have to pay the fee?

5. Your budget every week calls for saving $15 and spending $12. You buy some art supplies for what you think is going to be $11.50, but you forgot about sales tax. Will you stay within your budget if you buy everything, including the sales tax of 7%?

6. The stocks you bought for $40 are now worth $47. If you bought 32 stocks, how much more are they worth all together now?

7. Your credit card charges 24% annual interest. How much will the monthly interest be?

8. If you use your credit card to pay for something that costs $100, and you don't pay it off right away, how much interest will you end up paying the first month on top of the original cost?

9. If you don't pay off that $100 for an entire year, how much interest will you have paid at the end of the year?

10. If you travel to France, you'll need to change your American money into euros. When you land in Paris, you see a sign that says the exchange rate is 1 euro to $1.30 in U.S. dollars. If you change $50 into euros, how many euros will you get (provided there are no additional charges for the exchange)?

ANSWERS

1.

1. \$520 – \$104 = \$416 more
2. \$10 ÷ 4 = \$2.50
3. \$10 – \$2.50 – \$1 = \$6.50

2.

1. \$75
2. \$75
3. (4 weeks x \$2.50) + \$75 = \$85

3.

1. 3 hours x \$5 an hour = \$15, \$40 ÷ \$5 an hour = 8 hours
2. He has to make \$15 per week, so he would have to work for 3 hours.
3. 2 hours
4. 100 hours ÷ \$6 = 16 ⅔ hours.
5. He probably wouldn't have that much time because of school.

Hours worked for Customer #2	Money earned at \$6/hour
1	\$6
2	\$12
3	\$18

Hours worked for Customer #3	Money earned at \$12/hour
1	\$12
2	\$24
3	\$36

4.

1. $216
2. It is more, by $76.
3. $60
4. $200
5. $100
6. More, by $50
7. The second one, because she'll be able to keep a minimum amount of money in her account. She'll have to pay a lot of money to keep the first one.

5.

1. $.42
2. $141.27
3. $141.40
4. $501.50

6.

1. $29.99 + (2 x $15.99) + $34.99 = $96.96
2. 27.7
3. No, it's not more than half

7.

1. $253.94
2. Yes, she would have $103.94 left.

Date	Description	Withdrawal	Deposit	Balance
6/03	Clothes	$96.96	—	$253.04
6/15	Chore money	—	$20	$273.04
6/16	Ice cream cone	$3.50	—	$269.54
6/21	Headphones	$15.60	—	$253.94

8.

1. No.
2. Yes, she needs $4 more than she has now.
3. 60, 80, 100
4. $6.75
5. Yes ($6.75 – $20 < 0)

9.

1. $21.50
2. Yes, she has $1.50 left over.
3. No, she would need $10.50 more.

Description	Income	Expense	Available
Allowance	$10	—	$10
Allowance savings	—	$2.50	$7.50
Chores	$25	—	$32.50
Chore savings	—	$10	$22.50
Charity	—	$1	$21.50

10.

1. Yes
2. $24.96

Item	Price	Estimate	Actual cost
1 box spaghetti	$4.99 a box	$5	$4.99
2 tomatoes (each tomato weighs one quarter of a pound)	$1.19 a pound	$1	$.60
1 can tomato paste	$.89	$1	$.89
1 onion	$.60 each	$1	$.60
1 head garlic	$1.10 each	$1	$1.10
½ pound spinach	$3.99 pound	$2	$2.00
1 pepper	$.76 each	$1	$.76
2 carrots	$.54 each	$1	$1.08

续表

1 bottle Italian dressing	$2.75 per bottle	$3	$2.75
1 loaf French bread	$4.50	$5	$4.50
olive oil	$5.69 per bottle	$6	$5.69

11.

1. $1.39
2. $1.39
3. She spent more than her budget ($17.40 + $1.39 = $18.79).

12.

Day	Stock Price	Total Value of Shares
Monday	$76	$380
Tuesday	$79	$395
Wednesday	$61	$305
Thursday	$65	$325
Friday	$78	$390

1. $380
2. She made $10.
3. $90 x 5 = $450
4. $150 x 5 = $750
5. Three years from now, because the stocks will be worth the most money then.

13.

1. Yes, $40.10 more.
2. No, she will end up paying a lot more than $4,000 because of interest.
3. Yes, she would always stay under $4,000 still to be paid.

Month	Original Amount	Amount Paid	Interest	Still to Be Paid
1	$4,000	$30	0	$3,970
2	$3,970	$30	$39.70	$4,009.70
3	$4,009.70	$30	$40.10	$4049.80

14.

1. 619.50 pesos
2. 90 pesos x $0.08 = $7.20
3. (1−10%) x $50 = $45.00, $45 x 12.39 = 557.55 pesos

15.

1. 52 x $15 = $780
2. $15/3 = $5
3. Total amount = $600(1 + 0.005/12)^{12x1}
   Total amount = $603.01
   Interest = Total amount – original amount
   Interest = $603.01 – $600
   Interest = $3.01
4. No, you have more than $10 in your account ($68.50 – $45 + $36 = $59.50)
5. Not quite, you'll be $0.31 over (0.07 x $11.50 = $0.81, $11.50 + $0.81 = $12.31)
6. $7 x 32 = $224 more
7. 2%
8. $2
9. $24
10. 50 ÷ 1.30 = 38.46 euros